3RD EDITION

CHURCH IT
USING INFORMATION TECHNOLOGY
FOR THE MISSION OF THE CHURCH

NICK B. NICHOLAOU,
with **JONATHAN SMITH**

Copyright © 2024 Nick B. Nicholaou and Jonathan Smith.

All Rights Reserved.

No part of this book may be reproduced or transmitted in any form or by any means, electronic or mechanical, including photocopying, recording, or by any information storage and retrieval system, without permission in writing from the publisher.

Published by Church Law & Tax.
ChurchLawAndTax.com

Any Scripture taken from the HOLY BIBLE, NEW INTERNATIONAL VERSION is copyright © 1973, 1978, 1984 by International Bible Society. Used by permission of Zondervan Bible Publishers. All rights reserved.

All website references were current at the date of publication.

Editors: Matthew Branaugh and Rick Spruill
Art Director: Vasil Nazar
Illustrations: Rick Szuecs
Cover image: Gettyimages.com

ISBN: 979-8-9907785-5-9

Printed in the United States of America

So many churches and ministries have invited us to help them over the last thirty years! We are thankful to each, and dedicate this book to the bride of Christ, and to God's best gifts to us: our amazing wives and delightful kids.

Contents

Foreword 7
Introduction 11

SECTION ONE
Church IT's Mission 13
1. IT Department Structure 15
2. Who Is IT's Customer? 25
3. Leading in an IT Vacuum 35

SECTION TWO
Church IT Solutions 41
4. Selecting Solutions for the Wrong Reasons 43
5. Windows or macOS? 49
6. Church Management Software (ChMS) 55
7. Rightsizing Hardware 63
8. Virtual Computers 69
9. Software Charity Licensing 75
10. Making WiFi Work 81
11. VoIP 87

SECTION THREE
Church IT Strategies 95
12. IT Volunteers: Yes or No? 97
13. Training: The Most Neglected Spec 103
14. How to Keep IT Staff in the Local Church 109
15. IT Staff: Insource or Outsource? 117
16. Who Owns Your Public DNS Record? 125
17. Disaster Recovery and Business Continuity 129
18. The Twelve Months of IT 137
19. The Security Sweet Spot 145
20. The Value of Standardization 159

21. Changing Paradigms: The Cloud, BYOD,
    Social Media 165
22. Artificial Intelligence (AI) 177
23. Strategies for Solo Pastors 187

Glossary of Terms 191
Table of References 199
About the Authors 203

Foreword

In the 1980s, I was privileged to lead what is now called Christian Camp and Conference Association. Back then, every camping leader was scrambling to ensure the computer age didn't leave them in the dust.

Most of us had a cousin Eddie who gave us bad advice. I knew nothing, but I bought my first computer anyway—a Kaypro II. Few things in life are transformational. That machine was!

Fast forward to 1990 when I joined the management team at Willow Creek Community Church. The church had two IT guys then. The soft-spoken one installed a new computer in my office. With the bedside manner of a caring surgeon, he assured me my lack of technology savvy would not matter.

Gratefully, the church loaned us its IT team when we launched Willow Creek Association in 1992. We had an early version of an internal email system, but dozens of those pink "While You Were Out" phone message slips still ended up on my desk. We didn't leverage technology well.

During a Chicago snowstorm in early 1994, a headhunter from sunny California called me about leading Christian Management Association (now called Christian Leadership Alliance). I called my wife from the office and when I arrived home that evening, she was packed!

I'm sharing all of this to say, with deep appreciation, that when I arrived in Southern California, Nick Nicholaou was there to assure me that IT would be the *least* of my worries at CMA. *He was right!*

Nick's amazing firm, Ministry Business Services Inc., ensured that CMA—while seeking to model God-honoring management practices to thousands of church and ministry leaders—would *also* walk the talk and employ God-honoring IT practices in our national office and in our national conferences.

As CMA's national section leader for IT, Nick's mind and heart created a ripple effect of Kingdom fruitfulness.

Malcolm Gladwell's bestseller, *Outliers: The Story of Success,* notes the extraordinary power of the 10,000-hour rule. Gladwell would label Nick a "10,000-hour expert" based on the immense time Nick has invested with churches and ministries across the country. *He has seen it all—the good, the bad, and the ugly, yet he still has a contagious positivity about ministry.* Amazing!

So, without hesitation, I commend this very, very practical book to you. It is high time that a national leader elevated the foundational and critical work of IT! Nick gives you the information—and the why. His warm appreciation for the ministry of IT oozes out because of his high regard for the thousands of miracle-working IT people in the trenches.

So read and share this book with others: your pastor or boss, your colleagues, your volunteers, and your team members (if you have any!). Point out the helpful glossary in the back (a life-saver for me!) and the numerous resources.

When Rolla P. Huff was named president and CEO of EarthLink in 2007, *The Wall Street Journal* highlighted his priorities, which underlined the importance of an organization's operational side (including IT). Huff said,

> "It's all about execution.
> At the end of the day, I won't be judged
> on my plan as much as my execution."

Nick's insightful book will help you execute with excellence. And along the way, be sure to encourage and mentor others—like you—who are called by God to deploy the spiritual gift of administration.

John Pearson
Board Governance & Management Consultant
Author, *Mastering the Management Bucket*

Introduction

Why These Authors?

In his role as president of MBS Inc., Nick enjoyed the great privilege of serving Christian churches and ministries as an IT consultant and strategist. That started in 1987 after he and his wife, Grace, sensed God's call to serve the church in operational strategies. For thirty-five years he wrote hundreds of articles for print journals and contributed to many books. He was a national and regional conference speaker used by the Lord to help educate and bring awareness to wise IT strategies in the church. By helping literally thousands of churches through MBS, Nick helped the church navigate and establish strategies for the use of IT in ministry.

In 2006, Nick met Jonathan at a conference. In 2007, Nick felt the Lord's leading to invite Jonathan to join him behind podiums at conferences and to introduce him to his publishers. Over time, this collaboration led to the decision to set up Jonathan as the next owner of MBS, a transition that occurred in 2020.

Jonathan is an IT pro who is focused on ministry. He has served his church for years as its director of technology. During that time, the church has grown into a multisite megachurch, positioning Jonathan well to know and understand the IT trends, developments, and issues confronting churches and ministries. He is passionate about using technology for ministry effectiveness, and his insights further advance this book as it moves into its third edition.

The Purpose of This Book

This book is specific to IT in church and ministry environments. The church and ministry computing environment is unique. Through our work with churches of all sizes, we have seen what churches often do about IT issues that is *not* strategic. In the pages ahead, we address a number of the most common and costly IT mistakes churches make. We hope this book helps many church leaders think through better, more strategic approaches for applying computer technology to the missions of their churches—as unique as their computing environments may be.

Important Note

In the following pages we make recommendations for products we currently see as "sweet spot" solutions for churches (meaning an appropriate combination of capabilities while remaining sensitive to budget limitations many churches face). These recommendations come from an analytical process we continually work through. This means some of the hardware or software recommendations may change in the future, so keep that in mind. Use our recommendations as examples of an appropriate way to balance quality and budget range.

SECTION ONE

Church IT's Mission

When you have vision, it affects your attitude. Your attitude is optimistic rather than pessimistic. ... This is nothing more than having a strong belief in the power of God; having confidence in others around you who are in similar battles with you; and, yes, having confidence in yourself, by the grace of God.

Charles R. Swindoll [1]

---

[1] Charles Swindoll, *Dear Graduate: Letters of Wisdom from Charles R. Swindoll* (2007).

CHAPTER 1

IT Department Structure

> The work is too heavy for you; you cannot handle it alone. (NIV)
>
> **Jethro, Moses' father-in-law** [2]

There are five different technology "knowledge and skill" disciplines in most churches. Often church leadership thinks of them all as Information Technology (IT). That is probably because they all rely heavily on computer technology, and the assumption is that anyone *involved in* any of the five disciplines is equally capable of *serving in* any of the five areas.

However, each discipline uses different skills and tools. Unfortunately, churches often mistakenly group all five disciplines together into one "IT Department" and select a person strong in one of those disciplines to lead all five. It is one of the most common IT mistakes churches make.

It's similar to selecting the wrong doctor. Cardiologists and gynecologists are both important doctors. But their respective specialties are different and they are *not* interchangeable.

---

[2] Exodus 18:18.

The Five Disciplines of Church Technology

The five disciplines are (1) web and graphic design; (2) audio/video (A/V); (3) social media; (4) data infrastructure (the design and connecting of systems to ensure appropriate data flow at all levels); and (5) IT help desk. There are exceptions to each of these generalities, but the generalities reveal some of the dominant characteristics of the people who specialize in each IT discipline.

1. *Web and Graphic Design.* Wikipedia describes this discipline as "the process of visual communication and problem-solving through the use of typography, photography, and illustration."[3] It is similar to designing magazine ads that powerfully communicate a coordinated message all in one graphic, with the goal of moving the viewer toward an action. People who excel in this area are usually articulate communicators who are also very artistic. They use applications to draw, design, and do layout. They are often good project managers because of the timelines and project complexities involved in their work. However, they are usually not highly skilled in A/V or data infrastructure, for instance.

2. *Audio/Video.* A/V people are also creative communicators. They specialize in cameras, projection and lighting systems, soundboards, systems control boards, and storyboarding. They are also good project managers, so they can plan the A/V elements of a production from start to finish and make certain everything is ready on time. The computers used to render videos require a lot of resources and are often more powerful than some servers! Because A/V people use high-quality computers and are usually articulate communicators, they

---

[3] Wikipedia, https://en.wikipedia.org/wiki/Graphic_design (last visited Jan. 30, 2024).

are often tasked with IT oversight. But, like web and graphics people, they are usually not highly skilled in some of the other technology disciplines.

3. *Social Media*. Social media people are gifted communicators. They are highly skilled at leveraging today's electronic communication mediums (X (formerly Twitter), Instagram, Facebook, and so on) to start conversations with a wide audience and continue to engage them. While not highly skilled in web and graphic design, A/V, or in the data infrastructure required to connect them with the world, they are gifted in making appealing content that generates online interactions.

    Social media is often combined with web and graphic design or data infrastructure because people in those disciplines use social medial tools. But that often is a poor fit. It is best treated separately.

4. *Data Infrastructure*. Infrastructure people are more like engineers than creative types, and they are often not very good communicators. Their personalities tend to drive them toward specification analysis and the engineering of systems where the focus is to make certain data flows where it needs to go, whether that data is graphics, video, or data files (like spreadsheets). They tend to focus more on system designs, specs, and configurations, and are often not highly skilled in web and graphics, social media strategies, or A/V. They tend to be the introverts of the IT disciplines.

5. *IT Help Desk*. Help desk people are empathetic and great at patiently solving user problems. They are not usually as advanced in engineering as infrastructure people, but they do

a great job handling basic technology support needs. There is also a significant salary difference between help desk and infrastructure employees. While an engineer can handle "help desk" tasks, they are often not satisfied with them. At the same time, a help desk employee attempting engineering may be quickly frustrated. In ministry, the heart's desire to help matters, but that doesn't always make an infrastructure person a good fit for this role.

Tech Discipline	Typical Duties
Web & Graphic Design	• Creating websites (internal and external) • Graphic creation • Publication design, layout, and production
Audio/Video	• Operation and oversight of sound booth and related equipment • Video production (concept, storyboarding, production, editing)
Social Media	• Social media strategy formulation • Publishing and oversight of social media messaging
Data Infrastructure	• Analysis and engineering of system to support data transmission & storage • Configuration and support of computers and devices
IT Help Desk	• Empathetic and patient • Able to meet most users' support needs

Figure 1: Typical IT Discipline Duties

Which Discipline Is Best-suited to Oversee IT?

We are infrastructure guys, so our perspectives may be slightly biased. However, here is what we see at many churches:

- When non-infrastructure people oversee IT, the infrastructure requirements necessary to support the needs of the entire church staff are often underestimated. This is usually because of an inadequate understanding of infrastructure engineering and strategy and is evidenced by the tendency to buy and use non-enterprise hardware specs (meaning that the hardware chosen is usually not what corporations would consider appropriate to support high reliability needs with minimal cost). Less art than science, the data infrastructure discipline is more about engineering than creativity or communication.

- The infrastructure discipline is the foundational basis for all data transfer needs, so it usually is the best discipline to lead the technology needs of the church. However, infrastructure people struggle with communication. They also tend to take an over-restrictive approach to policies. These weaknesses can erode their relationships with staff and leadership (more on this in Chapter 3). Over-restrictive policies also often drive church staff to find workarounds, which can create numerous other risks and challenges.

Regardless of which discipline is responsible to lead a church's IT Department, here are some things that can be done to improve it:

- If a non-infrastructure person leads the IT Department, it is important to have a good and trusted infrastructure person in the department or to have a relationship with a good and trusted infrastructure consultancy. The infrastructure perspective is essential to having a system that works well for all staff and provides optimal and reliable transfer, storage, and backup of every kind of data needed by the entire church team.

- If the IT Department is led by an infrastructure person, that person should spend time getting to know the needs of the other four disciplines to make certain those needs are considered in the system design. The infrastructure person should also spend time with other church staff members and get to know their needs. In addition, the person will need a champion at the church leadership level to help overcome the communication and policy challenges mentioned earlier.

Because an infrastructure person needs someone at the leadership level to help overcome communication challenges and to implement policy changes, IT infrastructure is usually best placed under the chief operating officer (COO) or chief financial officer (CFO) of the church (or the comparable operational or financial roles if a church is too small for titles of that nature). Infrastructure is more of an operations discipline than it is a program discipline.

Communication Is Key

Many data infrastructure people struggle with communication. Jason Powell, though himself a skilled communicator as IT Director for Granger Community Church near South Bend, Indiana, says it is helpful to have a great communication skills mentor. Granger's IT Department is included as a part of its Communications Department, another good way to overcome the communications challenges.

David Brown, former Technology Director for Capital Christian Center in Sacramento, California, and now an engineer for Ministry Business Services Inc., a church and ministry IT consulting firm, agrees that communication is key:

> The chasm between vision and reality can be filled with jagged rocks. There has to be a bridge-builder who can communicate effectively in both worlds. The tech world can be too black-and-white or binary to communicate effectively to leadership. Being able to navigate necessary IT restrictions, while meeting the goals of leadership, will produce an outcome in which both sides are pleased.

Understanding IT Support

One of the challenges with managing technology in a ministry is understanding how IT support operates.

IT support is based on tiers (see figure 2).

	IT Support Roles
Tier 0 Self Support	Free, using own knowledge, Google, and asking other users
Tier 1 First Line Help Desk	Lower cost basic support for password resets, FAQs, etc.
Tier 2 Second Level Help Desk	Moderate cost in-depth support for complex issues that require detailed knowledge
Tier 3 Expert Support	More expensive expert help that may involve engineering
Tier 4 Vendor Support	Highest cost that is used when necessary

Figure 2: IT Support Roles

- *Tier 0* – often forgotten, but users can help themselves for free by using their existing knowledge and a browser search tool like Google. This is often what engineers use!

- *Tier 1* – This can be onsite or offsite. If it is offsite (for instance, MBS supports many ministries as their tier 1 support help desk), the onsite person may be a non-IT person who has technology abilities that an offsite support team can work with for basic tasks (such as help unjamming printers or plugging in computers). About forty percent (40%) of a church's tech help needs get resolved in tiers 0 and 1.

  Examples include: Help with a printer; a password reset; a problem with a monitor; an email reset; a name change; an account permissions change; or an issue that persists and requires escalation.

- *Tier 2* – Since this tier is only needed a little more than half of the time, why pay for it all the time by staffing it? This tier is for anything beyond Tiers 0 and 1, such as:

  A user's email works on their phone but not their computer; security patch management; software licensing and updates; complex issues that are not easily reproduceable; and system configurations.

- *Tier 3* – This is typically the last level handled on-site in a ministry, but it is often outsourced. This is only needed about thirty percent of the time. If Tiers 0, 1, and 2 can't resolve an issue, such as network traffic routing, firewalls, server infrastructure, core services, and custom code, then Tier 4 comes into play.

- *Tier 4* – This is an outsourced level of help. To make certain it's not a simple issue, Tier 4 usually starts at the beginning so engineering isn't involved trying to resolve something simple with an overly complex engineering fix.

The help desk role operates at Tier 1 or 2; the engineering role operates at Tier 3 or 4.

When it comes to staffing, do you have the right staff for the support you need?

CHAPTER 2

Who Is IT's Customer?

> Be a yardstick of quality. Some people aren't used to an environment where excellence is expected.
>
> **Steve Jobs** [4]

Many church team members complain that they feel like the IT Department restricts them. When a staff member expresses a need to the IT Department, the first answer is often, "No, you can't do that." This is even true when *leaders* express a need! IT restricts what can be done on church computers; for instance, IT won't allow software purchases, offsite data access, and more.

Some policies need to be in place to protect the church, but many IT policies are not a good fit for how church teams work. The challenge of finding a way to both protect the church and empower the team deserves an appropriate balance.

Why Is IT So Restrictive?

The answer is primarily an issue of influence. Church IT tends to be restrictive because of other IT environments that IT staff have worked in or been influenced by. IT in for-profit companies, academic insti-

---

[4] Jeffrey Young, *Steve Jobs: The Journey is the Reward* (1988).

tutions, government agencies, and healthcare are all very restrictive, the latter two especially so. People working in church IT are heavily influenced by their frame of reference—or their consultants' frame of reference.

In non-church settings, IT security is often very high and employees are seen as a commodity. When a non-church IT Department determines that a policy is important, violating that policy is grounds for termination.

Churches are different. Employees are not a commodity. Each employee—or *team member*—is someone we care about and for whom Christ died. In churches, we try to apply grace and mercy. And that can sometimes create a passive-aggressive culture where everyone perceives IT policies as nothing more than ineffective speed bumps that are okay to bypass—everyone, that is, except the IT Department! This passive-aggressive culture can also frustrate those staff members who are uncomfortable with ignoring policies. The IT Department can end up feeling unappreciated.

While consulting for a megachurch, Nick saw the passive-aggressive dynamic in an advanced state:

- The IT Department was frustrated that the rest of the church staff seemed to ignore many IT policies in unforeseen and uncontrollable ways.

- Many on the church staff bragged to me about how they got around IT restrictions. Even top pastoral leaders in the church bragged about this. The executive pastor encouraged his team members to come up with ways to get past restrictive IT policies so they could get their jobs done!

- Some of the more compliant staff were frustrated on behalf of the IT Department and could not understand why everyone treated IT so disrespectfully.

Who Is IT's Customer?

One of the best ways to overcome this common conflict is for the IT Department to approach its role as though it is a for-profit business and every member of the church team is a customer.

Why does that help so much? A for-profit business knows it must meet its customers' needs effectively, competitively, and in such a positive way that the customer will want to come back and do business again. Anything less threatens a for-profit's very existence. Imagine the effect this underlying philosophy would have on most church IT Departments—and on the church teams they serve!

That means IT must not dictate what people may or may not do on the system, *except when it is absolutely necessary to protect the team members and/or the organization.* Restrictive policies that are not truly necessary have the same effect as crying wolf when no wolf is there: people stop paying attention to the warnings. Thus, any restrictive policy needs to be carefully scrutinized to determine how essential it is before being put in place. Top church leadership can help determine the validity of the policy and require buy-in from the team. Leadership's response to the policy, and a decision to enact it, can be the force behind it, rather than the force coming from the IT Department.

Who Sets Policy?

IT teams often feel pressured to set appropriate IT policies to protect the organization, its team members, and its data. But that is *not* IT's responsibility. Rather, organizational leadership should set policy—with IT giving its input consultatively. IT's responsibility, in turn, focuses on implementing the policies leadership establishes.

For every IT policy need, IT should meet with organizational leaders and let them know that a policy is needed and why. IT needs to effectively communicate the need, the risks, the alternatives, and the costs in non-technical terms.

For example, IT may want to implement something called two-factor authentication to improve security for systems and devices.

- If IT were to set that policy, those on the ministry team who didn't like the new policy (probably almost everyone!) would complain to leadership that IT is being too paranoid and heavy-handed.

- *If*, however, IT met with leadership and explained—in terms everyone could comfortably grasp—the reasons for two-factor authentication and the risks of not implementing it, and then let leadership make the decision, then the team's angst would be with leadership, not with IT. Risk management is leadership's responsibility, so the decision is best made by organizational leaders!

This approach offers at least two benefits:

1. It helps end the tendency of some team members who try to get around IT's policies. These are *not IT's policies*—they're leadership's policies!

2. If leadership decides to not set a certain policy, and a problem later arises, IT will be there to help resolve the problem, but

IT also will know it did its job when it properly advised organizational leaders.

We'll develop this concept further in Chapter 5.

Say "Yes" Whenever Possible

Clif Guy, Church of the Resurrection's IT Director in Leawood, Kansas, once told us about his IT team's astonishing approach to requests: his church IT Department's policy is to say "yes" whenever possible. This approach is rare. But it is exactly the way IT should run in the church. The default posture always should be to say "yes!" That does not mean IT always says "yes," though. Sometimes the response really needs to be "no, and here's why." But having a "yes" default posture helps convey that IT is on the same team as the rest who are driven to fulfill their call.

If the request was for something the IT team is uncomfortable with, or even unsure of, then saying so is okay with the promise to fully respond within one or two business days—again, though, with the hope of being able to say yes. To do this well, IT needs to understand the need, and how the staff member hopes the request will meet it.

Most church team members rarely understand the results of some IT-related decisions on overall productivity. IT needs to say "no" on occasion, but IT needs to say "no" so infrequently that when IT does, the rest of the team will heed it.

Here Are a Couple of Practical Examples

1. Local Admin

One big debate is over something called "local admin." When setting up a team member's Windows computer, the option exists to give that team member administrative authority over his or her computer. Doing so does not affect the security rights the team member will have on the network and its data, but only on the computer the team member uses day in and day out. It only affects the computer, and not the overall system, so it is referred to as local admin rights.

About half of church IT people believe in giving everyone local admin rights, and the other half adamantly oppose it. We are big proponents of giving all team members local admin rights unless someone proves he or she just cannot safely handle that much authority. Giving everyone local admin rights offers some valuable benefits:

- It allows team members to install printer drivers, flash drive drivers, and so on, without having to contact the IT Department.

- It allows the user to run application updates and patches without the IT Department's intervention.

- It dramatically reduces the number of support calls to the IT Department, lowering the time—and potentially, staff—needed to handle the load of resolving all those support calls.

- It builds "customer" satisfaction.

However, there is a risk to giving everyone local admin rights. For example, a team member can load malware on a system or delete a necessary operating system file. After using this strategy on thousands of church computers for many years, we have found this risk to be minimal.

If your church grants local admin rights to all team members, IT needs to engineer ways to protect and resolve issues that could arise because of local admin rights. However, doing so is a significantly lower load and cost than *not* giving local admin rights. Here are some ways for IT to protect the system:

- Fully back up the church's data daily, and keep at least three to four weeks of those backups in case a problem isn't noticed right away.

- Utilize imaging and deployment tools offered by Microsoft and Apple to maintain current images of computers so that reestablishing a computer's configuration can be quick and easy. This can be further simplified by buying similarly spec'd computers in bunches rather than individually, since all similar computers can be backed up by a single image.

If a team member repeatedly shows he or she cannot be trusted with local admin rights, change only that team member's rights instead of restricting everyone on the team. This helps IT build a level of "customer satisfaction" that is a blessing for every member of the team.

2. Convergence of IT and A/V

The Audio/Video discipline is converging with IT, and a common denominator—data infrastructure—is the reason. A/V is beginning to depend more heavily on infrastructure technology—specifically using Internet Protocol (IP) over Ethernet cable and fiber for communication purposes. This forward shift in A/V communications potentially offers big cost-savings. It also requires the two disciplines to work together, something IT team members and A/V staff are often reluctant to do.

IP over Ethernet and Fiber

For nearly four decades, IP over Ethernet cable, and more recently (about two decades) fiber-optic cable, have been IT's primary ways of moving network data. Because of this, IT has a lot of experience on how to manage and maximize these mediums of digital communication.

The problem, from the A/V perspective, is that IT is not willing to bring this experience to bear on A/V needs. The reason: A/V needs two specific things turned on to make its system perform correctly—things that IT has been told not to turn on:

- Internet Group Management Protocol (IGMP); and
- Jumbo frames (also known as "jumbos").

Over the decades, IT has learned that the more that is turned on in the network, the more complex and less reliable the network becomes. Normal data networks do not need IGMP or jumbos, so IT leaves them off.

But A/V needs IGMP and jumbos.

Given A/V's needs and IT's reluctance, A/V people conclude the A/V system must run through its own separate network, including separate cable and switches. A/V needs IGMP and jumbos, and since IT says "No," A/V needs budget funding for that separate network.

How Can These Two IT Disciplines Get Along?

Here's where IT can play a vital role in helping all members of the ministry team fulfill their calling—while also saving the ministry thousands of dollars.

Recent and current IT infrastructure can certainly handle the weight of the A/V team's needs. IT may need to do some research to learn how best to configure the system, but it boils down to adding a number of VLANs[5] running on higher-quality switches with things like IGMP and jumbos turned on. The IGMP and jumbos can be restricted to VLANs only used by A/V. This restriction enables A/V to rely upon the IGMP and jumbos it needs to do video multicasts. It also allows IT to protect the network's reliability.

If IT isn't willing to help A/V in this way, then A/V will have to buy several higher-end switches to meet its needs. It would be better to focus the funds A/V would unnecessarily spend to instead improve the reliability and quality provided by the better switches for all IP communications (which are needed all week by the entire team, not just A/V's needs six to eight hours per week) than to have to maintain physically separated or duplicated networks.

This Is Not Rocket Science!

Okay, maybe it is. But the point is that infrastructure folks divide digital communications over the same cable into multiple virtual networks all the time! With the right network equipment (enterprise-quality firewalls, switches, and cable that may already be in use at your church), different kinds of traffic over the same cable can coexist without interfering with each other. They can be set to act as though they are on separate cables, even though they're not.

That could mean that A/V doesn't need a separate network. And not requiring a separate A/V network can save a lot of money for implementation and for support over time.

The key is in the quality of the network hardware and its engineering.

---

[5] VLANs are *Virtual* Local Area Networks—very common in data networking. They function as though they were separate cable networks, but actually all run on one cable (thus the term Virtual Local Area Networks).

Not All Network Equipment is Created Equal

Though we'll develop this topic further in Chapter 7, buying "right" really makes a difference. It's sort of like Goldilocks and the Three Bears. You can buy network hardware that is:

- Underpowered (consumer specs),
- Overpowered (probably very expensive), or
- Just right! (effective mix of features and price)

The basis of A/V horror stories about sending A/V communications in a VLAN over the data network almost always points to inadequate hardware or engineering.

And this is where converging the two IT disciplines of A/V and infrastructure can prove to be wise. Your infrastructure team has been working with IP communications based on decades of experience and knowledge. IT needs to learn how to meet A/V's needs and then agree to help, and A/V needs to trust IT to meet its needs.

CHAPTER 3

Leading in an IT Vacuum

> If it wasn't hard, everyone would do it. It's
> the *hard* that makes it great.
>
> **Tom Hanks** [6]

Church leadership makes decisions and sets direction for ministry programs, but usually without input or feedback from the IT Department. Yet most decisions made today involve IT disciplines. Why does this happen? How do we reasonably overcome it?

The Challenge

Ministers, boards, and committees meet regularly to talk through the status of their church and identify strategic directions to increase their effectiveness in their communities, among other things. These strategic directions might be about seasonal programming (Easter and Christmas, for example), group programming (young adults and families, for example), or community event involvement (such as a local parade). Each of these touch upon an IT discipline (graphics, A/V, social media, and/or infrastructure).

Unfortunately, IT leaders often learn of these programming plans late in the process.

---

[6] *A League of Their Own,* Columbia Pictures (1992).

Common examples we have seen in churches include:

- "I just found out that we're supposed to provide WiFi to the entire congregation in one week for something the pastor will do during his message."

- "We're supposed to provide check-in stations in a new part of the building and add a new kiosk in the lobby, but we don't have the bandwidth or connections in that part of the building or a secure connection in the lobby. I only have four days to make this work."

- "We added a pastor to the staff today, and we need to give the new pastor a computer."

We also receive calls about A/V and website issues. The problem in these examples is that IT wasn't told early enough to adequately prepare for the situations, yet they are responsible for making them work! Worse, when the situations involve a program or event that has been already promoted, there's no way to adjust the schedule to allow for good preparation!

Very few churches and ministries include someone responsible for IT in the discussions and decisions for programming. Many churches see IT as operations overhead, so IT is not typically represented in those planning meetings.

The Consequences

Two common consequences of not having IT involved early enough in the planning process are:

1. The programming decisions made stretch beyond the current capabilities of the church's IT disciplines to fulfill without

expanding hardware, software, or talent. This is not inherently negative, but it can be if the cost is higher than leadership anticipated.

An example might be the decision to add financial transaction terminals to the campus (giving kiosks, credit card terminals, and so on). Even if credit and debit card numbers will not be maintained locally, transmitting them over the church's network and internet connection can trigger something called PCI Compliance, which can be very expensive depending on how the terminals are connected. Strategic implementation requires advance notice.

2. Expansion of IT hardware, software, or talent cannot be strategic when advance notice isn't given. We have seen many churches and ministries use hardware that is nowhere near "enterprise grade" quality. Instead, they have "consumer grade" solutions in place.

   Consumer grade solutions often get implemented because IT is up against a tight deadline and has to look for something—*anything!*—available in a store immediately. In Chapter 7 on "Rightsizing Hardware," we discuss the basis of a common IT industry catchphrase: "If you can find it on a shelf in a store, you probably don't want it!" When under a very tight, last-minute deadline, on-the-shelf consumer grade solutions may offer the only hope of avoiding failure. The result is low-quality solutions, which often cost more than appropriate solutions if given the appropriate time to order. From there, the problem persists because instead of replacing lower-quality solutions once the deadline passes, the church keeps the consumer-grade solutions in place. That often translates into long-term reliance on solutions that have higher support costs and poorer performance—and that lowers team productivity.

What Does Corporate America Do?

Larger companies often have a CXO position (this is a general reference to a top leadership level person where the X is a wildcard; **C**hief **E**xecutive **O**fficer - CEO, **C**hief **O**perations **O**fficer - COO, **C**hief **F**inancial **O**fficer - CFO, and so on) who is responsible for all things IT. Although many churches and ministries are smaller than those large companies, we can still learn from this approach.

Many medium and larger corporations have a Chief Information Officer (CIO) or Chief Technology Officer (CTO) who participates in leadership meetings and serves on boards. These individuals help shape decisions and decision-making processes so IT can get the right solutions in place to facilitate plans.

What is a Reasonable *Church* Solution?

Adding a CIO or CTO to church leadership teams isn't the solution. Churches are focused on reaching the lost for Christ and want as much of their budgets as possible to go toward programming rather than overhead.

IT usually reports to the CFO- or COO-equivalent in a church. That person often has the title of church administrator, church business manager, or executive pastor. Regardless of the title at your church, the person in that position needs to represent IT in leadership meetings in the same way a CIO or CTO would. The problem is that church CFOs and COOs usually don't know a lot about IT—especially infrastructure. That's okay if IT is managed strategically. We recommend the following:

- If leadership meetings happen regularly (weekly or monthly, for example), the CFO or COO should also have a regularly scheduled meeting with the IT person who reports to them to talk about the plans and directions that are being discussed in the leadership meetings *immediately* following the leadership meetings.

- If these follow-up meetings happen right on the heels of the leadership meetings, the leadership planning discussion will be fresh in the mind of the CFO or COO. He or she can then get the IT person's input and relay any concerns to leadership within an appropriate timeframe. This can help leadership adjust planning if needed, and it gives IT notice of what may be coming so that IT can research and plan accordingly.

This is not an IT power grab! Rather, these meetings will help leadership accomplish plans for less money and will give IT enough advance notice to strategically research the appropriate enterprise-class solutions necessary for upcoming plans.

SECTION TWO

Church IT Solutions

> Computers allow us to squeeze the most out of everything.
>
> **Buzz Aldrin** [7]

---

[7] Buzz Aldrin Tells Moon Landing Story, *Red Carpet News TV* (Aug. 3, 2012), http://redcarpetnewstv.com/buzz-aldrin-tells-moon-landing-story/ (last visited Jan. 30, 2024).

CHAPTER 4

Selecting Solutions for the Wrong Reasons

> People say I make strange choices, but they're not strange for me.
>
> **Johnny Depp** [8]

When deciding how to meet a specific need, churches frequently ask other churches about the solutions they used to meet that need. Whether it's hardware, software, or a database, churches like to ask other churches what they use and whether they're happy with the choice to use it. That kind of research is good in theory but may not be in practice. It can have some significant downsides.

The Quest to Be a Good Steward

Churches have IT budget constraints because they want to focus as much of their budgets as possible on programming. So, the faster and more efficiently they can find operational solutions, the better.

Right?

---

[8] Johnny Depp Fan Page, Facebook (July 12, 2012), https://www.facebook.com/TheJohnnyDeppPage/ (last visited Jan. 30, 2024).

Sometimes. In these types of situations, churches often turn to pastors and leaders from other churches for referrals. These referrals can help us avoid mistakes and point us in directions that are good, but they can also point us in directions that are bad.

Referral Request Assumptions

When we ask other churches for referrals of IT solutions, we make certain assumptions that may or may not be true. These assumptions are worth discussing. Here are two of the most significant:

- *Assumption 1:* "We are nearly identical churches in the way we do ministry, the things we value and prioritize, who we target in fulfilling our mission, the way our staffs are structured and work together, and the way we minister to our congregation. Our two churches are nearly identical in personality, style, target, and focus."

- *Assumption 2:* "The other church researched its choice well. Leaders there did their due diligence in searching out the best possible solution based on the church's personality, style, target, and focus (which are just like ours!); the search was objective and exhaustive, landing on the best solution. In asking another church's leaders what they recommend, we save time and energy by not duplicating their research."

Problems with These Assumptions

With the first assumption, it is unlikely the two churches are identical. Although many churches have similar goals of reaching the lost and helping them become true followers of Christ, the ways each church accomplishes its goals are different. These differences are shaped by the personalities, education, skills, and gifting of the staffs and leadership. The mission of a church shapes it, affecting its strategies to reach people. The location, the surrounding community, and the physical layout of a church also shapes it. Each church has a unique personality, style, and culture—so unique, in fact, that it makes it unlikely that the two churches are identical.

That said, it is unlikely one church wants to do the exact same thing as the other. We frequently see this when churches are looking for new Church Management Software (ChMS). When one church relies heavily on another church or individual's recommendation, the solution rarely lasts five years. The dissatisfaction is often expressed in statements like, "The software just doesn't match our church's ministry needs and style."

In the second assumption, the problem is that the referring church's leaders probably did *not* do an exhaustive search for the best solution to match their needs. Rather, they probably did the same thing you would probably like to do—call another church and ask them for a referral. If you trace the chain of referrals, you may find that the first church chose the solution simply because of a discount made possible by a friend or a spouse, rather than the "best of class" solutions available for its unique needs. Even if that decision somehow worked out for the first church, it is unlikely the same solution will work for the subsequent churches using its guidance.

Ultimately, decisions made solely or based too heavily on the recommendation of another church are likely to fail.

A Better Way

The best method is to do an objective, exhaustive analysis using the unique needs of your church to decide what solution is best. However, this requires a significant investment of time, and someone to see it through.

The best way to determine your church's needs is through surveys and/or interviews with your team members most affected by the solution. The key to success is to do this phase of research without an agenda favoring any of the ministries at your church.

From there, all available solutions should be reviewed to determine which one best matches your church's needs.

Nick did this kind of research for churches up until he retired. His method is a bit more "old school." When it came to evaluating ChMS options for a church, he would interview each team member (sometimes in individual interviews and sometimes in focus group interviews). In each interview, he would first establish the person's role to understand that person's perspective. Then he'd ask a few questions:

- What software and hardware do you use in your job? (This helps identify how many silos of data exist.)

- What tasks does the system do for you now that you consider crucial? (This helps document tasks the new system must be able to do.)

- What does the system do that slows you down or gets in the way? (This helps document potential areas of improvement the team would like in their new solution.)

- How could the system serve you better?

- What else would you like the system to do for you?

You may notice he would never ask what solution the person wished for! The interviews identify needs that must be met in search of the best solution.

Nick asked these questions in interviews (rather than through a questionnaire) because some of the answers would lead to follow-up questions that gleaned information he would otherwise never get. Then he compiled a customized survey to send to all ChMS providers to help determine which solutions would best meet the church's needs. He also sent a modified copy of the survey to the team members so they could weight the importance of each item on the survey with choices of unimportant, might be nice, important, essential, or no opinion. He multiplied the numeric values of the results of both groups to get an objective score for the providers most likely to meet the church team members' needs well.

Once you have objectively identified the top three solutions, you and your team then should schedule live demos or visit other churches using them. This step helps your team members determine whether the solutions are good to use the ways they would want to use them.

By involving your larger church team in the process, you will also help overcome one of the chief complaints of church staff. In an article by church leader Chuck Lawless, he recounted the way staff members often feel excluded from decision-making processes, making them feel they have no voice—that their opinions do not matter.

Giving your team a voice in the process takes more time and energy, but it increases team buy-in, encourages team members that their opinions are valued, and gets everyone focused on the positive aspects of the selected solution.

Do Your Research. It's Worth the Effort!

Quick decisions many times are made because such-and-such church uses a particular solution. Checking with other churches can help shorten the list of preferred solution providers. It also may add some solutions to the list that you may not have been aware of! Don't make the decision solely based on those recommendations.

Do the research with your staff (they will thank you!). Gather a list of criteria to use in the decision-making process, and either require live demos, visit churches with the solution in place—or both—to see if what your church chooses will fit your church's culture.

CHAPTER 5

Windows or macOS?

If you don't mind, it doesn't matter.
Anonymous [9]

Church leaders debate whether to run the Windows operating system, the macOS operating system (which requires an Apple Mac)—or both. This debate has caused angst for years. Let's address it.

We researched the pros and cons of churches running only one or both systems many times over the years, and formed an opinion based on the compatibility of the Mac processors with other computers on a given church's network.

In its early days Apple focused on Motorola PowerPC processors to drive its Macs. These processors didn't coexist well on networks using Intel x86 processors. We therefore recommended churches focus on one platform or the other, but not mix the two.

Then, in 2005, Steve Jobs shook up the computing world when he announced Apple was transitioning Macs to Intel x86 processors! When we refreshed our Mac testing a couple of years after the

---

[9] Anonymous government researcher referenced in *Statesville Record and Landmark,* Statesville, N.C. 1968.

announcement, we concluded Macs finally worked well on Intel-based networks.

Because of these developments, we no longer believe a mixed-platform environment is a problem. Churches may use both Mac and Windows operating systems on the same network without compromising performance. Macs tend to cost more, which can affect church budgets, but outside of that consideration, we see both platforms as equally good.

What Really Matters

Salaries and benefits paid to staff are the most expensive budget item for most churches, so the efficient productivity of each team member is crucial to the mission of the church.

If one staff member will be most efficient on a Windows computer, while another staff member will be most efficient on a Mac, churches should consider giving each one the platform that maximizes the staff member's efficiency. Doing so amplifies the value of the cost spent on the person's salary and benefits package, and doing so boosts the efficient fulfillment of the church's mission.

Sharing Files and Collaboration

Some church leaders wonder whether mixed platforms will create problems with sharing files and collaboration between staff members. Most software solutions today run on both platforms with only minor differences in their feature sets, so sharing files and collaboration can work well.

Are There Dangers?

Yes, there are! But with a little strategy, providing both platforms to your church team can be done safely.

- Mac enthusiasts strongly believe a few things, even though the facts say otherwise. Let's discuss four of the most common assertions made about Macs:

    1. *They are more expensive.* Just compare similar computer specs for both types of operating systems. Again, though, if someone will be more productive on a Mac, it may be worth the additional expense.

    2. *Macs require less support.* This one is a little more complex to address. Because Mac users are often independent in their approaches to technology, they help each other, and this can lower the number of support issues your IT Department or vendor must address.

        However, hardware failures with Macs are as high as those running Windows systems, and they often require making an appointment for face-to-face time with a Tier 1 support person (cleverly branded as a "genius") at a local Apple store. This is far more costly than the next-day warranties of some other manufacturers (like Dell) who will have someone come to your church to fix the computer.

    3. *Mac users often prefer "better" solutions.* That independent spirit mentioned above means that Mac users sometimes will prefer using apps that are different than—and incompatible with—those used by Windows users.

    4. *Macs are safer from malware and more secure.* Totally false! The Mac OS is easier to hack than any other operating system.

Each of these four assertions can be managed and effectively addressed. Here's how:

1. Mitigate the higher cost of Macs by focusing on the efficiency and productivity gained by the staff members who prefer using Macs.

2. Look for a vendor in your area that is an authorized Mac repair vendor. The vendor may offer more personalized and convenient hardware support. If they're approved by Apple for performing warranty repairs, then warranty repairs will be covered as though they were directly handled through Apple—and non-warranty repairs may cost less!

3. Ask church leaders to develop and adopt a policy regarding which software programs—like Microsoft Office as an example—the team should use. This will reduce—or even eliminate—the possibility of staff members trying to share incompatible files.

4. Require all Macs to run anti-malware software (we recommend SentinelOne (best purchased through a discounted cost arrangement exclusively through MBS.[10]). Mac users also should be reminded regularly to physically protect their Macs at all times (for instance, not leaving them on a table at a coffee shop while they use the restroom), particularly if they are using MacBooks.

After completing our updated Mac research in 2007, Nick published an article[11] and mentioned some of the weaknesses of the platform like

---

[10] https://www.mbsinc.com/solutions-apps/
[11] Nick Nicholaou, *Can I Get a Mac?*, MBS, Inc. (Jan. 1, 2008), https://www.mbsinc.com/can-i-get-a-mac/ (last visited Jan. 31, 2024)

we did here. Mac enthusiasts came unglued! He got more email from that article than almost all others combined. So he wrote a follow-up article factually explaining each of the weaknesses mentioned in the first article, and didn't get even one email.

After many years of using a Windows-based computer, Nick has used a Mac almost exclusively since 2007. The bottom line for church leaders is this: whether it is a Mac operating system or a Windows operating system, the two platforms are good network citizens. It's okay to let team members use whichever platform works best for them—with the stipulation they use leadership-approved software to create their files.

CHAPTER 6

Church Management Software (ChMS)

> What turns me on about the digital age,
> what excites me personally, is that you have
> closed the gap between dreaming and doing.
>
> **Bono** [12]

Using computers in churches is different from most other settings. How IT Departments in churches serve their "customers," and which solutions churches choose to use, also are different than most other settings.

These differences are apparent numerous ways, but most especially in the software used in church offices.

In the mid-1980s, Nick did the first exhaustive study on the software used by churches to manage financial and non-financial data. At that time, there were 262 solutions available. He established some reasonable minimal criteria for review, and only 17 made it through that filter! That meant there were 250 inadequate solutions available to churches! With so many to choose from, it was nearly impossible

---

[12] Paul David Hewson (Bono), (2005) TED Conference.

for a church to do a good and thorough analysis itself. Unfortunately, that meant many churches chose poor solutions.

Since then, many solution providers have merged, consolidated, or gone out of business. Many new solutions have sprung up, too!

We collectively refer to these solutions as Church Management Software (ChMS). Today, about 75 viable solutions created specifically for churches exist. That's still a lot to choose from—but our work researching these options is designed to help your church make a wise choice.

ChMS Landscape

The category of ChMS software has changed in recent years. Most who currently offer ChMS solutions are for-profit entities who are driven by the transaction processing fees that come through online giving. Based on this recent history, keep in mind that a ChMS partner you have today could be bought, sold, or merge with another company in the future.

Most churches already have a ChMS in place. Both real and perceived dissatisfaction with their software often pushes many churches to think about choosing a different ChMS. Before considering a change, though, make sure you've exhausted your current software's training resources and that you have a deep understanding of it. Don't make a biased decision to change, even if another church recommends a different solution that seems poised to address problems your church has encountered with its current solution.

No ChMS is perfect, and knowing the inevitable financial costs one brings (including productivity loss), you don't want to rush a change only to find you swapped one set of problems for another because your choice wasn't well-researched.

If you've evaluated the situation and decided that it's time to make a change and you feel prepared for the costs and the lost staff productivity, then it's time to do some homework. The objective evaluation of all the software options is on the technology decision-making process mentioned in chapter 4.

Why ChMS Instead of More General Solutions?

The financial and non-financial needs of churches are unique and not easily met with non-ChMS solutions.

In the financial area:

- Church accounting systems need to track various income and expense transactions, many of which close to different "capital"-equivalent accounts that are restricted (temporarily or permanently) or unrestricted. This is an accounting issue that sets church accounting needs apart from most other organizations:

    - Income and expense accounts close to a capital account—in companies it might be retained earnings; in churches it is a fund balance, also referred to as a net asset.

    - The unique aspect of nonprofit accounting is that income and expense accounts tied to various purposes and funds (like the missionary fund, the building fund, and so on) need to close to those various fund balance accounts and produce an available balance of how much is in each fund being tracked.

This helps fulfill one of the unique responsibilities churches have in their accounting systems, which, according to the Evangelical Council for Financial Accountability (ECFA) is

> using the gifts on a timely basis within the limits of the giver's restrictions. To accomplish this, ministries must have accounting systems which track not only the restricted gifts in the [ChMS] but also track the expenditure of the funds and unused funds (net assets with donor restrictions). . . Some [ChMS] and accounting systems do this better than others. The more giver-restricted gifts your ministry receives, the more important it is to have systems that efficiently handle these gifts.[13]

- Many churches keep their books on a "cash basis" or "modified-cash basis," with only a small percentage of churches on the "accrual basis." A ChMS solution needs to accommodate all three. (Modified-cash basis is a mixture of the cash and accrual methods in which some items, such as accounts payable, for example, are managed on a cash basis.)

- There are unique aspects to managing payroll in churches. For example, ministers' income may or may not be subject to income tax withholding based on individual elections. Additionally, the housing allowance portion of a minister's income designated and approved by a church needs to be treated as non-taxable for federal income tax purposes, and none of a minister's income is subject to FICA (Social Security and Medicare) withholding.

---

[13] *Four Accounting Challenges for Ministries* ECFA (2023), https://www.ecfa.org/Content/Four-Accounting-Challenges-for-Ministries (last visited Jan. 31, 2024).

In the non-financial area:

- **Demographics and contact information.** Churches track all kinds of demographic and contact information about people. That's not uncommon among Customer Relationship Management (CRM) solutions. What makes it much more complex in churches is the tracking of numerous other activities. These include contributions, attendance, background checks for serving in children's ministry, event registrations, life events, and family relationships. .

- **Children's ministry.** An important database function in many churches is the ability to securely check children in and out for children's ministry events, such as Sunday school. In addition to the check-in and check-out processes, churches need to track allergies, restraining orders, restricted access, and many other pieces of data that keep children safe—and help parents feel their children are safe.

- **Attendance.** Tracking attendance helps identify those who may need follow-up and encouragement.

- **Giving.** Tracking contributions and pledges is essential. This feature is normally considered a ChMS non-financial feature because, although many solution providers do not offer their own accounting solution, they still need to offer contributions tracking. These systems must satisfy the need for the church to document contributions for donors who may wish to seek—depending upon their tax situations—charitable contribution deductions when they file their income taxes.

Reasonable Expectations

Churches and ministries often ask for help in looking for a new ChMS database and/or accounting system. During those conversations, we talk through what their team can expect from whatever new solution they choose. Most people don't realize what "there is no perfect solution" truly means. The very best anyone can expect is an 80-percent to 85-percent match with the expectations and needs of any church or ministry. Stated from the other side of that equation, the lowest number of missed expectations they can hope for is about 15 percent to 20 percent.

If a church already has a top-tier solution, the process of searching for and transitioning to a new solution may be nothing more than trading one 20-percent set of missed expectations for a different set of 20-percent missed expectations. If that's true, then that church may be better off exploring other ways to overcome their dissatisfaction before starting a transition. We'll address this further in Chapter 13.

Data Normalization

In nearly every ChMS consultation, we tell the church that, based on the old adage "garbage in/garbage out," the church's data must be cleaned up before importing the data into a new ChMS database. The industry term for this process is "data normalization."

Data normalization is needed to create reliable, efficient uses of the ChMS. It becomes a need because a church team either fails to establish a set of standards before any data gets entered—or fails to train people on the standards set, including new hires and volunteers. Without correction, the problem only grows larger.

Here's a common scenario that illustrates this problem:

One existing data point for the children's ministry may be the name of the school each child attends. Children's ministry staff and volunteers may complain about the unpredictable results they receive whenever they search for all of the kids from a particular elementary school. The cause? Different team members have entered varied spellings for each school, either because no standard was set or there wasn't training with an established standard. For instance, Abraham Lincoln Elementary School may have been entered into the ChMS as:

- Abraham Lincoln Elementary School
- Abe Lincoln Elementary
- Abe Lincoln
- ALE
- Possible misspelled variances

There's no way to get a predictable list when searching for a data point that is not normalized.

The solution to fix this is very tedious. But it is necessary. Someone needs to look for every record with this data point and normalize it to the standard the church wants to use.

Think of how many different data points your church has in its database—the task of normalizing all of them seems insurmountable! But a normalized database is essential. It will produce better, more predictable results with the ChMS in the long run. Such a project offers a terrific opportunity to use volunteers, so consider ways to recruit and oversee them to do it.

CHAPTER 7

Rightsizing Hardware

> It is far more important to be able to hit the target than it is to haggle over who makes a weapon or who pulls a trigger.
>
> **Dwight D. Eisenhower** [14]

One of the most common mistakes churches and ministries make is related to the hardware they buy.

The mistake manifests itself three ways:

1. Incredible overspending on highly sophisticated hardware the church will never take advantage of (overbuying),
2. Buying consumer grade hardware that doesn't work very well in a corporate network setting (underbuying), and
3. Buying hardware built locally.

In each of these mistakes, the church usually ends up spending more on its purchases than necessary, either in outright purchase costs or in higher support costs and lost staff productivity.

---

[14] President Dwight Eisenhower, Address to the American Society of Newspaper Editors and the International Press Institute, (April 17, 1958).

We work hard to stay neutral with hardware, software, and platforms so we can effectively evaluate the options in the marketplace and make appropriate recommendations. This chapter reflects that neutrality.

Overbuying

Most vendors do not understand the needs of churches, and thus have a difficult time making appropriate recommendations. Some overestimate the church's needs, recommending hardware with sophisticated features the church will never use.

Overbuying is often the result of one of three things:

1. *Buying based on possible future needs.* Many churches think somewhat aggressively when expressing their growing technology needs. But if those anticipated needs don't materialize for three or four years, the result will be overbuying.

2. *Recommendations based on limited understanding.* Churches are unique in their hardware needs. If the church trusts the counsel of someone who has limited perspective (someone whose IT expertise is focused in another industry), the result often will be overbuying.

3. *Vendor recommendations that affect the vendor's profit.* We have been to many churches that previously used IT vendors who sold them hardware based on what they made, sold, or had in stock, but the hardware didn't come close to appropriately meeting the church's needs. It could be because a member of the congregation worked for the company selling the hardware. In many cases, we have concluded the overbuying by a church

is often caused by a vendor trying to get a larger sale, rather than trying to meet a church's specific need.

The most common example of overbuying occurs with some high-end firewalls and network switches. Though powerful, these feature-rich solutions can almost definitely meet any church's needs; in fact, they significantly *exceed* the needs of most churches! The problem is that the features of those high-end firewalls and switches require a deep understanding of IT communication technology, and so their configuration and maintenance require a certified engineer. That is expensive! And those higher-end features rarely benefit churches and ministries.

A few years ago, a church needed a local area network (LAN) revamp because its current configuration wasn't meeting its needs. When looking at its server configuration, we discovered the server surpassed many found in high-end datacenters! The church spent more than twice what it needed on that server, and the church would never come close to taking advantage of its huge processor, random-access memory (RAM), and storage capacities.

Underbuying

Many churches and ministries mistakenly buy their computers from big-box retailers, electronics stores, or via internet searches. Many IT professionals have a saying in common: *If you can find what you need on a shelf in a store, you probably don't want it.* That goes for computers, printers, switches, WiFi routers, and more—almost anything IT.

There are differences between enterprise-spec systems and consumer-grade systems. Here are a couple that may not be readily apparent:

- The engineering and match-up of components are at a higher level in enterprise systems than are usually found in consumer-grade or gamer systems. What many don't realize is that most computer components for both specs (consumer and enterprise) are manufactured on the same assembly line or process. The components are tested after being manufactured, and depending on how they score in testing, their performance results will determine their model number and associated cost. Thus, similar-looking components can be in both systems, but they will perform very differently over time. Enterprise-spec systems usually include higher-rated components and have a higher level of research and development (R&D) invested in them by the company whose name is on the product (Dell, HP, and so on).

- Consumer-grade computers usually have a lot of trialware (software you get to try before you buy, but probably don't want). Companies offer trialware to try to influence you to buy their software after trying it. This approach is like advertising and helps underwrite the cost of the computer, which lowers its purchase price. Trialware usually means a less-efficient user experience, and less-efficient user experiences translate to wasted time—and sometimes higher support costs.

  Another problem with trialware is that it often doesn't uninstall completely. The little bits and pieces left behind can sometimes be the cause of problems that users later encounter.

Underbuying is often the result of having to buy something fast because there wasn't a strategy in place to anticipate needs and allow purchases to be planned. Planned purchases usually mean better stewardship because they allow time for ordering enterprise-spec systems at the lowest possible cost. Enterprise-spec systems don't necessarily cost a lot more; they're just better in an enterprise (non-consumer) environment where systems are networked.

Buying Locally Built Systems

In the early days of personal computers, many people thought buying locally built computers was best. They thought the support would be better since they were built nearby. Maybe in the 1980s that was sometimes true, but more often, *locally built systems need more support* because they do not have the benefit of the level of R&D a larger company, such as Apple, Dell, or HP, could invest in them. Even consumer-class systems have some R&D. But locally built systems are only *assembled*. A local shop may do its best to buy great components, but local shops have limited opportunity or capability to test how each of those components work together. Local shops certainly do not have the buying power to request manufacturer firmware changes to components that would improve the way their component interacts with other selected components.

Surprisingly, some churches still buy locally built systems! There is no way a local shop can compete with the quality, reliability, and support of a well-engineered system. And another surprise: locally built systems almost always cost more, not less!

So buy right! Buy enterprise systems for your church or ministry. You will save in the end.

Discovering What to Buy

Finding out what is best to buy can be challenging. Stores can't help you buy enterprise-spec hardware, so where can you turn?

The best answer is to find a vendor with expertise in configuring, engineering, and supporting enterprise networks—preferably in church and ministry settings. At the end of Chapter 15 we list a few resources that can help, but that list is not by any means exhaustive.

There are also some association resources that can help you find someone that can help:

- The Church Network (thechurchnetwork.com);
- Church IT Network (churchitnetwork.com);
- Christian Leadership Alliance (christianleadershipalliance.org); and,
- Your denominational, movement, or association office.

CHAPTER 8

Virtual Computers

> It's better to be a fake somebody
> than a real nobody.
>
> **Matt Damon, portraying Tom Ripley** [15]

In the late 1990s, a technology was introduced into modern computing that previously existed only on large mainframe and midrange computers. That technology, called virtualization, made it possible for a computer or server to concurrently run multiple computers—on one hardware device!

At first, we were skeptical. How would it affect performance and reliability? And why would anyone want to do such a thing?

For those churches aware of this technology, many ask the same questions. In reality, though, most churches aren't aware. At national and regional conferences, when we ask for a show of hands, rarely do even 25 percent of church leaders indicate they understand what virtualized servers are.

---

[15] *The Talented Mr. Ripley,* Miramax Films (1999).

This technology can save churches and ministries a lot of money, improve network reliability, and improve network administration. And it's free! Well, free for most churches, anyway—more on that in a bit.

How Does Virtualization Work?

Virtualization requires an app called a *hypervisor,* which functions like a shim or wedge between the bare hardware and the operating system (like Windows, macOS, or Linux). Rather than installing the OS directly onto the computer or server, the hypervisor is installed first. Once it is installed, a wizard then helps you install as many virtual computers onto the physical computer or server as it has resources (processors, RAM, storage, and so on) to host! Thus, the physical computer is often referred to as a host. To help explain how this works, the steps to creating a host and the computers and servers that will run on it are:

1. Install the hypervisor on a computer *before* installing any other operating systems (the hard drive should be empty when installing it).

2. Use the hypervisor's wizard, or menu, to say you want to create a virtual computer or server. Answer the hypervisor's questions about:

    a. What operating system you'll install on the virtual machine.
    b. How many processors you want the virtual system to have.
    c. How much RAM you want the virtual system to have.
    d. How much storage (hard drive size) you want the virtual system to have.

3. The hypervisor then takes a few minutes to create the virtual computer as you've spec'd it *logically* (versus *physically*). Once it's done, it will tell you to insert the OS installation media (like inserting a DVD or pointing to an ISO file[16]). From that point forward, it looks like you're doing a normal OS installation, except that it's running in a window because it is a *virtual* computer!

Depending on the host's physical resources, you can install many virtual computers and servers. In my firm's datacenter, we run dozens on a single host!

Why Would You Want a Virtual Computer?

Physical computers and servers cost a lot of money. Surprisingly, their physical resources go largely unused for the most part. For instance, if you were to monitored the processor activity level in a server after it starts up, you'd find that it rarely goes above 10 percent utilization after the server's start-up. The processor is the most expensive component in the machine, and yet most of the time it is nearly idle! The same is true for many other resources. Virtualization helps maximize the return on investment (ROI) made when purchasing that machine.

When we first began working with virtual servers and computers we were cautiously skeptical that a virtual server could deliver the high degree of reliability churches need. We learned a properly configured one could.

One reason virtualization improves reliability: the best way to set up network servers is to do so with each major separate network service having its own server. So, an email server should run on its own server, as should a database server, file server, endpoint server, and

---

[16] An ISO file is an image of a CD or DVD that can be mounted logically just as though it was physical media. You can also burn an ISO file to optical media and run it that way.

so on. That means the best reliability requires having many servers. That strategy would be out of reach for most church budgets were it not for virtualization.

With virtualization, you can have a couple of physical servers—or hosts (we call them vh1, vh2, and so on, for virtual host 1, virtual host 2, and so on)—provide the resources for many virtual servers (see Figure 3). That amounts to large savings—with improved reliability. A win-win!

Virtualization is what powers the cloud! All of those cloud solution providers are running nearly all of their servers as virtual servers. Doing so saves money on hardware and substantially improves reliability

Brands and Costs of Hypervisors

There are two main players in the hypervisor marketplace: VMware's vSphere Hypervisor and Microsoft's Hyper-V. Both are free for most churches.

VMware pioneered the technology, and its solution, vSphere, is currently the most mature, powerful, and the easiest to use. Hyper-V is good, but vSphere has the edge, which is why we recommend it.

Since hypervisors are free for most churches, what is the business model behind them? These companies make their income through tools and features that only large organizations need. The best way to tell if your church needs those features is to determine whether or not your organization uses a Storage Area Network (SAN) on the network. SANs are *very* expensive devices, usually costing more than $25,000. If you have a SAN, you know it because of the high cost, and the relatively small cost of the hypervisor features you would want would be a slam-dunk decision to purchase.

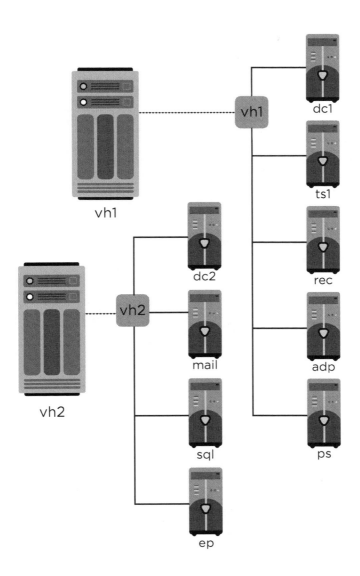

Figure 3: Two hosts running nine servers

Most churches do not have a SAN. That means the basic set of features included in the free hypervisor should take care of your needs!

Virtualization in the Cloud

Using virtualization, cloud-based providers like Amazon, Microsoft, and MBS can offer you your own server at a fraction of the price of attempting to host that server, or a virtualized host, on your own.

Not only has virtualization changed how servers are hosted on your network or on your premises, but it also empowers the cloud to provide server hosting at reasonable prices.

The Bottom Line

The bottom line on virtualization for any church with one or more on-site servers is that if your church hasn't already, it most likely should be taking advantage of the power and reliability virtualization can add to your network. The price tag is certainly no barrier for doing so!

CHAPTER 9

Software Charity Licensing

> I was called to do this, and when you are called to do something you don't give up. If you believe in something strongly enough, you continue to push until you find a way to make it work.
>
> **Chris Booth** [17]

In the early 1990s, Chris Booth and Nick (and probably others, too) independently lobbied solution providers to offer licensing discounts to churches and ministries similar to those available to academic organizations and government agencies. The first large provider to agree was Novell, a company that pioneered the software that made standardized local area networks (LANs) possible. Then, mostly due to Chris' efforts, the second was WordPerfect, the pioneer for word processing software. Microsoft Corp. was third. (Microsoft eventually won both of those software categories due to its code development focus and market strategy.)

---

[17] L. Roberts, *A Higher Calling: Chris Booth gave up corporate life to bring bargain high-tech supplies, services to nonprofit organizations,* The Journal Times (Nov. 28, 2008), https://journaltimes.com/lifestyles/faith-and-values/religion/a-higher-calling-chris-booth-gave-up-corporate-life-to/article_e2d23fbf-310c-515c-83fa-db7e4124c478.html (last visited Feb.1, 2024).

Today, many solution providers offer discounted charity licensing discounts at, or near to, academic licensing discounts. Surprisingly, we still meet many in ministry who are unaware of these terrific discounts!

Thus, they unnecessarily overspend on software.

Why Are Charity Discounts Offered?

Discounts are a great way for solution providers to offer philanthropic assistance to organizations that help society—but there is also a solid business case for doing so!

Board members and many other volunteers run churches and ministries. Their hearts are definitely tied to helping the churches and ministries they serve to be effective.

Many of those people are decision makers or decision influencers in their paid jobs. If they become familiar with church computing products and solutions that are effective and reasonably priced, they often end up championing those same options at their paid jobs. Discounts to churches and ministries make for very inexpensive marketing for the solution providers!

Lee Iacocca, an auto industry leader who saved Chrysler from near bankruptcy, used a similar strategy. It was one of his many successes. In his autobiography[18] he said the challenge in the auto industry was to find new ways to get the car-buying public to test drive Chrysler's cars. When he realized people actually paid to test drive cars when they rent them, he made selling new cars at a reduced price to rental car companies a priority. That translated into many test drives!

---

[18] Lee Iacocca, *Iacocca: An Autobiography* (1984).

Where Can You Get Charity License Pricing?

Many software sellers can sell charity licenses. The key is to ask your software solution vendor to make certain they are using charity-licensing SKUs when selling your church its software. If you're not certain that you are being offered true charity licensing, check with the solution provider (Microsoft, VMware, and so on). If a solution you want to use doesn't offer charity-licensing discounts, ask the vendor if it is willing to extend its academic licensing discount to your organization. Really, it is in the vendor's best interest to do so. Don't pay full retail because doing so is usually not necessary for a charity.

Charity Licensing Pitfalls

With these discounts comes a caution.

A few companies are changing the way they offer charity-licensing agreements, and it involves a danger for many churches and ministries. These solution providers include specific restrictions in the charity license language that sometimes conflict with the doctrine of many churches and ministries. Avoiding those licensing agreements may be in your church's best interest.

The two most prominent solution providers including these restrictions are Google and Microsoft. Both do it exclusively with their online solution charity-licensing agreements, but—interestingly—not with their academic licensing agreements. In Microsoft's case, it also is not doing this with its volume license charity-licensing agreements.

In the charity-licensing agreement for Google's G-Suite, charities are required to certify they do not discriminate in hiring or in employment practices related to the lesbian, gay, bisexual, transgender, and queer (LGBTQ) community in any way. Microsoft's Office 365 charity license is similar. However, Nick successfully convinced Microsoft to add both a statement and a FAQ explaining how churches and religious organizations may not be required to adhere to its certification requirements based on state and federal laws that govern churches and religious organizations and allow certain types of discrimination.

So, Microsoft is making room for churches and ministries in its licensing agreements. Google, however, does not.

This is not a legal issue (though it could impact a legal proceeding). Rather, it is a mechanism Google is using to filter which organizations it wants to donate or discount software to. As a company this is Google's right, and churches and ministries shouldn't falsely certify any positions they hold just to receive a discount or donation.

In raising this concern, we are not taking the position that positions like Google's should or should not be an issue. Christian churches and ministries love individuals in the LGBTQ communities. They want to serve them by introducing them to Christ and helping them grow in their relationships with him—value-changing relationships!

However, for many organizations, the employment practices requirement in the agreements like Google's pose a problem. It's important for churches and ministries to be aware of these restrictions and respond as they feel led.

Don't Be Afraid of Hearing "No"

One book Nick read while working in the auto industry dealt with negotiating deals, and the author made a point that stuck with him: most people don't ask for concessions when they have an opportunity to negotiate because they're afraid to hear "no."!

From that point forward, Nick began asking for concessions more often. Sometimes one gets a "no" response, but many times, one will get a "yes" if the request is reasonable.

So, if a vendor does not offer charity-licensing discounts yet, ask for it. The worst the provider can say is "no"!

CHAPTER 10

Making WiFi Work

Whenever you find yourself on the side of
the majority, it is time to reform.

Mark Twain [19]

Many churches today broadcast WiFi on their campuses for the convenience of staff and guests. WiFi seems to behave mysteriously, though, and only a small percentage are completely pleased with how their WiFi works. There are reasons for poorer-than-expected performance (largely related to hardware specs). Additionally, there are strategies every church should use to protect the church and to protect those who use its WiFi.

Can WiFi Be Reliable?

The short answer is *yes*, but the long answer is *yes, if you understand how it works and engineer accordingly*. The problems many experience with their WiFi campus deployments are fairly predictable, and easily avoided.

The most common WiFi complaints include:

---

[19] Mark Twain, *Mark Twain at Large* (1878).

- *People can't get a connection.* This is most often a hardware spec problem. The cause is usually from using consumer-spec WiFi access points (WAPs), which directly relates to the discussion in the "Underbuying" section of Chapter 7 (see page 65). The WAPs you find on a store shelf and buy usually have a limit of about 25 connections. Some think that buying a bunch of them will overcome that limitation, but their engineering can't and doesn't.

- *The WiFi is too slow and devices drop their connection.* WAPs broadcast their signals over channels that, if not managed, bleed over and interfere with each other. Also, few routers or firewalls have the appropriate specs to accommodate the large burst of internet traffic that happens during church services and programs.

  Remember, the needed engineering and configurations for WiFi used during church services are similar to those needed for a small (or maybe not-so-small) conference at a convention center! And just as it would be true for a conference at a convention center, the use of this resource is usually low—until the crowd shows up! That's when the WiFi must be set up appropriately to meet the needs of that crowd.

- *People can access content we don't want them accessing.* WAPs that are simply plugged into church network switches or routers can allow access to the internal network systems (like servers and printers) and to inappropriate websites. They need to be managed by a firewall or controller, something few off-the-shelf WAPs can accommodate.

Hardware

We mentioned that off-the-shelf WAPs are limited to connecting about 25 devices at a time. Buying ten of them does not increase the capacity tenfold to 250 as you might anticipate. The problem is that their hardware doesn't have the feature set necessary to allow them to work together. Instead, they conflict with each other and even interfere with each other.

We learned this the hard way. We found the answer was to recommend WAPs designed to operate in a commercial environment—such as one in a convention center. We prefer Ruckus WAPs. Ruckus has some WAPs that can accommodate hundreds or more concurrent connections and can coordinate with other WAPs when there are many deployed on a campus.

That is different than "mesh" technology, which doesn't handle the larger requirements of church guest WiFi needs well.

Firewall

The next piece of the solution is the firewall. In Chapter 20 we explain why we recommend SonicWALL firewalls. The key to a good WiFi system is that the firewall needs to be able to handle the volume of internet traffic that happens during church services and programs. As we mentioned, few off-the-shelf units have what's needed.

The firewall also helps protect those connected to the church's WiFi from websites that have inappropriate content. This is essential—especially if children might be connecting to the internet.

Configuration

When you search for a WiFi signal with a device you'd like to connect (smartphone, tablet, computer, and so on), you see a listing of names. Those WiFi signal names are Service Set IDentifiers, or SSIDs. We typically configure three or more SSIDs for churches and these are managed in the firewall. They are an important part of a good WiFi strategy.

The three SSIDs we usually configure are

1. *Guest*. This SSID only has access to the internet, and that access is filtered in both directions (incoming and outgoing) to prevent adult-oriented website access.

2. *Lay*. This SSID is similar to the Guest SSID, but it is password-protected and has protected bandwidth to accommodate lay leaders who need to access YouTube or other similar streaming content while teaching a class. It might also include access to a printer to accommodate printing notes or handouts.

3. *Staff*. This SSID is password-protected and may even be hidden so that only those who need it know it exists! It is only for church staff and includes access to the internet and to network servers and devices.

WiFi Security

It is wise and prudent for churches to protect access to their WiFi system. The best recommendation from a legal perspective is to password-protect each SSID.

A church in Missouri learned the hard way that open and unprotected WiFi can cause significant problems. The church's WiFi was not protected with a password, and it ran around the clock. Someone drove into the church parking lot after hours, connected to its WiFi from the parking lot, and distributed child pornography. When the FBI investigated, two things happened that hurt the church:

1. The FBI confiscated the church's computers—*all* of them, including the servers! The FBI confiscated them because it is required to do a forensic examination to determine if anyone on site was involved with distributing child pornography. The church was without its computers for months, a major disruption given how much the church depends upon computers for operations, study prep, and so on.

2. Television and newspapers covered the story. A headline in one newspaper was "Child porn investigation focused on [church name]." A 'film at eleven' type of TV news report was titled "Child porn linked to church IP address." You can imagine how this bad press impacted the church in the community, and how long it takes to recover from the impact.

Most churches, in the name of being a *welcoming church*, don't want to password protect their guest WiFi. For churches who make that choice, we recommend—at a minimum—they turn off their SSIDs that are not password-protected when no church service or program is running. That's only possible (without manually unplugging all WAPs) if the church is running an enterprise-class solution like the one MBS recommends (SonicWALL firewall with Ruckus WAPs). The SonicWALL can turn the guest WiFi on and off based on a set schedule that the IT manager can override as needed.

Taking Security a Step Further

While considering WiFi security, let me mention a vulnerability I see at many organizations (not just churches): default passwords still in effect for devices. This applies to WAPs, but it applies to other IT devices, too. A good practice is to always replace the default password on *any* IT device.

Gary Messmer, MBS's lead engineer, was visiting an auto dealership for routine car maintenance. While sitting in the waiting room, he looked to see if there was a WiFi signal (SSID) he could use. He found one, but it was password-protected. The SSID was named the default name with the WAP model assigned by the WAP's consumer-class manufacturer! Being familiar with those devices, he decided to try the default password. He was in! He connected to the dealership's database and pulled up someone's record, walked over to the service manager, and while showing him his computer screen asked, "Is this your dealership database?" The service manager was shocked!

In Chapter 20, we discuss why churches need to be proactive about managing their IT security. All data is vulnerable, but in the church, we need to make certain we are doing our due diligence to protect our data and the information that could hurt someone if it were inappropriately accessed.

CHAPTER 11

VoIP

> The overall point is that new technology will not necessarily replace old technology, but it will date it. By definition, eventually, it will replace it.
>
> **Steve Jobs** [20]

Several years ago, we decided MBS needed a new phone system. Our system at that time was a peer-to-peer system, and it was significantly limited.

Our Challenge

When looking at our options, we were concerned with the prices normally associated with new phone systems. Typical PBX phone systems (Private Branch Exchange, typified by having an operator or receptionist) can cost tens of thousands of dollars!

To serve churches and ministries well, MBS runs on a very thin margin, so we didn't want unnecessary expenses. We looked at less traditional options that wouldn't cost as much. Most churches also operate on thin margins, so we realized that the fruit of our quest might prove useful to churches as well.

---

[20] Steve Jobs, *Steve Jobs: His own words and wisdom,* (2011).

Initially, the solution seemed to be Voice Over Internet Protocol (VoIP), but even many of those systems cost more than $10,000. At a Church IT Network event Nick attended around that time, several attendees discussed a free PBX server that could run as a virtual server. He was leery because "free" is often more expensive than it's really worth. But considering the expense of the alternative, he decided to research this further and give it a try.

There are two aspects to VoIP:

1. Connecting your building to the world telephone system via VoIP technology versus standard phone line technology (in Figure 4 below, this is referring to how the server in the bottom-right corner of the church connects to the worldwide telephone system).

    - This is what most people think of when they think of a VoIP system.

    - Standard phone lines are costly, and VoIP connections can save you significant money! (MBS provides SIP for churches that includes unlimited trunks and usage. There are no metered lines and no trunk limits.)

2. The internal corporate phone system, known as a PBX, that connects church team members via extensions. (In Figure 4 below, this refers to all the phones and the computer they connect to inside the church.) The PBX is figuratively the server in the bottom-right corner of the church.

Figure 4: Illustration of VoIP Components

PBX Technology

The PBX concept and its technology is very mature (dating back to the 1800s!), and there now are many free, reliable, open-source PBXes available. At the Church IT Network event, many recommended one open-source PBX, so Nick decided to test it to see if it was as good as suggested.

MBS downloaded a solution called Asterisk and configured it as a virtual server on one of its hosts. It was solid but challenging. A colleague who works on VoIP PBX systems recommended we switch to a version of the same solution called FreePBX. He said it was based on Asterisk, but with more features and a better graphical user interface (GUI). MBS switched and hasn't looked back!

That Got Us Thinking...

One of MBS's goals is to help Christian churches and ministries focus as much of their budget as possible on their ministry programs. Since churches and ministries could save a lot of money by making their next PBX upgrade a FreePBX server, MBS needed to help them move in that direction! It started helping churches and ministries implement FreePBX phone systems because our conclusion agreed with the initial research: PBX solutions are a very mature technology, and spending tens of thousands of dollars on a phone system just isn't as necessary today as it was in the late 1900s!

Options Are The Biggest Challenge!

Designing a VoIP phone system can be overwhelming because the technology is so flexible and capable. It has a lot of options. Here are some to consider:

- *Physical Handsets or Softphones with Headsets?*

    - One option is to eliminate physical handsets (the traditional telephones that sit on a desk) and instead use softphones. Softphones are apps that run on computers, tablets, and smartphones. Some of the benefits include:

        - *More desk space for other things.* This is similar to how no longer using large CRT monitors allows for more desk space.

        - *A hands-free environment.* Softphones depend upon headsets. Most users love that!

- *The ability to interface with the computer or device's contact list, such as Microsoft's Outlook.* Using softphone apps means the softphones can interface with your contact list.

- *Saving money.* Buying softphones is economical, costing less than handsets.

- *Less Ethernet ports needed.* Using softphones eliminates the need for having two Ethernet ports at each desk (optimal configuration for desks with a computer and a handset)—more on that later.

* *Live Attendant or Automated Attendant?* Many churches still want a live person to answer phones rather than an automated attendant. VoIP PBXes give you multiple options, which are:

    – Live attendant;

    – An automated attendant greeting that gives options for service times and locations and a list of departments; it also encourages the caller to enter an extension if known, or to use a dial-by-name directory; or

    – A live attendant backed up by an automated attendant when the live attendant isn't logged in or isn't able to answer within a specified number of rings.

* *Hosted Off-Site?* Having your VoIP hosted in the cloud by a reputable provider can ensure 99.999-percent uptime for your church's voice service. That's less than 20 minutes of downtime per year. When hosted in a datacenter, phone communications

can still happen even in a power outage by using an internet connection somewhere or running a smartphone softphone app.

VoIP phones, whether handsets or softphones, log into the PBX server. Because they are IP phones and connections, they can login from anywhere they have an internet connection!

This feature serves mobile workforces in the church well. For instance, if pastors like to do their sermon prep from home, they can login their handset (if they have one at home) or softphone to the PBX and it's like they are just down the hall at the church office. You can simply call them on their regular extension.

Also, when someone changes offices, they can easily keep their extension without calling a technician to program the system.

Using a softphone app on your Android or iPhone to call someone shields the cell number in the caller ID, and instead sends the church's caller ID. This is helpful to protect church team members during off-hours because it keeps everyone they call from learning their personal cell number.

One of our clients is a missions agency, and this aspect serves them very well. The agency's overseas office logs into the same hosted PBX as the home office, so the offices on both continents are simply extensions to each other. When team members travel overseas, they have the softphone app on their smartphone log into the PBX, and they are, again, simply an extension. They really like being able to get an outside line in the United States, no matter their location. And there are no international dialing issues or long-distance rates!

Consider how helpful this feature would be for short-term mission trips. We recommend having a few extensions and softphone apps in reserve and ready to assign to short-term mission trip leaders. While on the trip, if leaders have WiFi access on their smartphones, they can securely call the church by simply dialing an extension—and can likewise receive calls!

Future Proof

There is no doubt the COVID-19 pandemic changed the world, and VoIP was no exception. While it is impossible to be 100-percent future proof, the biggest hinderance when churches and ministries had to work remotely was call management.

Churches that already embraced a VoIP solution quickly pivoted. Churches still running traditional PBX systems struggled to manage phone calls when no one was at the physical church location to answer the phone.

While no one knows what's next, VoIP provides churches and ministries with maximum ministry flexibility.

Important VoIP Warnings

VoIP options may appeal to churches. They have their advantages, but they can also pose challenges. Several ways VoIP can cause problems include:

1. *Two Ethernet ports are required.* MBS is known for data networks, and has had many clients who were previously sold a poorly set-up VoIP system from a phone company. The challenge is that most churches do not have Ethernet ports at each desk (VoIP phones connect to the church network with an Ethernet cable just as computers do). Using a modern VoIP phone,

you can plug your computer into the phone, and the phone into the network without any performance issues. But you still need an Ethernet drop for each physical phone.

2. *VoIP PBXes can be vulnerable to hacking.* It is important to protect the PBX properly by putting it behind a capable firewall. Make certain you keep the PBX software and the server's OS current. Many of the patches released have security enhancements that are important to have in place.

SECTION THREE

Church IT Strategies

> Trying to predict the future is like trying to drive down a country road at night with no lights while looking out the back window.
>
> **Peter F. Drucker** [21]

---

[21] Peter Drucker. Origin unknown.

CHAPTER 12

IT Volunteers: Yes or No?

> We have different gifts, according to the grace given to each of us. If your gift is ...
>
> **The Apostle Paul** [22]

For many reasons, churches thrive when their congregations are involved in the work of the church. We love—and need—our volunteers! They are crucial to churches.

What should you do with congregants who want to volunteer to help with your IT?

There are ways IT volunteers can help and there are ways they can hurt.

Let's address some common issues that cause problems for church teams when volunteers are involved. The goal here is to help church IT leaders effectively use volunteers and avoid common missteps in using those volunteers.

Someone who volunteers for a church IT team should understand the church's perspective and priorities so that their help is most effective.

---

[22] Romans 12:6.

What Can Go Wrong?

The IT system needs to be protected from the possible whims or mistakes of a volunteer. Try walking into your local bank and volunteering to clean the vault. When folks volunteer to help with your church's IT, be careful to not simply give them the keys to the vault.

When churches rely on IT volunteers, here are a few things that can go wrong:

1. Poor IT strategy. Churches often suffer from poor IT strategies because they allow people—volunteers and staff—to learn and develop their IT skills while serving the church in this vital area. Letting people learn and grow in the church IT Department sounds good, but the truth is that it is usually only good for the one learning and growing.

While these individuals are improving their knowledge and skills, they are doing so by trial and error because that's one of the ways people learn. The unfortunate result is a less-than-optimal IT strategy and solution, leading to unnecessary break downs.

2. Fragmented IT strategy. This is a compounding of the first point. When asked to evaluate or help with IT strategies, people bring the perspective of their experience. If that experience is different from those who came before—and it nearly always is—then any new strategy introduced will be based on the experience and perspective of the newest team member. That usually results in a fragmented IT strategy that doesn't mesh well with other strategies that are in place, affecting staff productivity.

It will be magnified further if the perspectives invited to weigh in on the IT strategy have limited church IT experience.

We saw this play out at a megachurch where a new congregant, who was a sales engineer for a large hardware manufacturer, introduced himself to one of the pastors after the service. When the pastor learned of the new congregant's employer, he invited him to evaluate the church's IT strategy. When the congregant did, he noted that the church experienced higher-than-normal reliability with zero downtime, but said its strategy wasn't *scalable*. (Scalability refers to the ability of a system to scale up to an extremely large configuration, like when a company goes global. It is often used by salespersons to encourage sweeping IT changes.) He recommended that the system be redone with scalability in mind.

The pastor agreed that scalability must be important since it is a church with global reach, so a plan and budget was created using equipment from the congregant's company. The budget was **huge**, and the size of the church ten years later is about the same as it was when the sales engineer—using a term few outside of IT fully understand—sold the church on the need to be scalable. It was terrible to see so much money wasted.

3. Lack of availability. These poor and fragmented strategy issues mean increased support needs. Volunteers generally are not available when they are needed most: when church staff are at work. Volunteers are usually only available before or after work hours. More productivity is lost because problems can't be resolved until after hours. Although the church staff usually absorbs this for a while, the dynamic often creates conflict.

4. Burnout. The problem comes full circle. Poor, fragmented IT strategies require more and more after-hours support from the volunteer, who eventually concludes that a choice must be made between supporting the IT strategies that he or she helped put into place or spending time with friends, family, and activities. If the volunteer pulls back or bows out, the church then must find another person to take over.

5. The cycle repeats, And so the cycle starts again, and things will likely get worse by further IT strategy fragmentation.

Church IT Complexity

The IT needs of churches are more sophisticated in their complexity than most realize! Here's why:

- *Sophisticated complexity*. Rather than a simple list of names and contact information, churches track many dimensions of congregants' lives, including contact information, family structures, life events, interests and spiritual gifts, volunteerism, background check data for certain volunteer roles (like children's workers, for example, to ensure no pedophiles are involved), contributions, attendance, and more. In addition, churches use many audio/video solutions (hardware and software), and they need to run with the excellence and reliability of a broadcast studio.

- *Mission critical*. Church staffs are usually smaller in number than their for-profit counterparts and have less computer application training, but they still have deadlines that must be met every week. Dan Bishop with Houston CO-OP in Texas once said the church is more like a newspaper business with daily and weekly publication deadlines than almost any other business analogy. He is correct! Only a reliable system can relieve the level of stress these deadlines create for smaller teams—every week, on time, every time.

- *Budget sensitivity*. Churches with small teams, working with daily and weekly deadlines, also have limited budgets. Therefore, IT strategies need to tackle sophisticated complexity and mission critical reliability as efficiently as possible. Anything less will hinder the mission of the church.

The Right Place for IT Volunteers

A few professional IT vendors specialize in helping churches meet IT needs well. At the time of this writing, we know of four or five—two of whom have a national reach. At a minimum, they can help set the overall IT strategy in your church and identify roles in which volunteers can help.

Some possible volunteer IT roles that can be successful include:

- Help desk, *if* they're available during weekday work hours or on weekends to provide needed off-time for IT staff (see Chapter 4);

- Pulling cable;

- Moving equipment;

- Deploying/installing software (if given a step-by-step list to help achieve standardization among systems); and

- Some periodic maintenance routines, such as cleaning keyboards, mice, and monitors/displays. More on this in chapter 18, "The Twelve Months of IT."

- Another way churches involve IT volunteers is to help teach those in the community how to use technology. Community classes on social media, smart phones, and basics of Microsoft Office can be great outreach tools and a good use of IT volunteers. Jonathan's church, Faith Church, does this at least twice a year.

Each of these is helpful and important, and inviting volunteers to help in these ways will be a blessing for staff members and volunteers alike.

CHAPTER 13

Training: The Most Neglected Spec

> Organized learning must become a lifelong process.
>
> **Peter F. Drucker** [23]

In Chapter 4, we said the best way to choose new software solutions is by conducting a thorough analysis of needs followed by an objective search (see the section titled "A Better Way," beginning on page 45). When using this approach, the staff will recognize that no software solution is perfect and focus on 80 percent to 85 percent of the needs the new solution will meet. Staff will clean up their data, receive training, and begin enjoying the blessings of the transition.

Then normal change and growth begin to affect the staff. New people join the team through growth and attrition, but the new people were not part of that software search process. They never had to go through the pain of cleaning up old data and might have never been trained.

Two things begin to happen: (1) newcomers use the software incorrectly, degrading its effectiveness and the integrity of the data; and (2) as they struggle to understand how the software works,

---

[23] Peter Drucker, *Classic Drucker* (2006).

they look for alternative solutions that are simpler for their or their department's needs. These dynamics cause the entire staff's focus to begin to shift from the vast majority that works well for the entire church to the small percentage that doesn't. Often, the church ends up concluding the current software doesn't work anymore, even though it mostly does. Lack of ongoing training is really at the heart of the problem.

Training is the Key

Whether it is ChMS or another technological tool used, prioritize ongoing training for your teams. Few IT investments will yield better returns.

The two areas of IT that most affect the entire staff are ChMS and productivity tools. Here are the strategies we recommend:

- *ChMS*. Require on-site training by the solution provider for all staff when implementing a new system. Require on-site training every year, too.

  This will:

  - Give your team members the ability to discuss their "pain points," so the trainer can either show ways to improve processes or communicate those pain points to the solution provider. Caution the staff not to fall into an unwillingness to change a process because "we've always done it that way."

  - Encourage your team to learn about new features the solution offers and emphasize how these features can help fulfill your ministry vision. It's essential that the solution continues to grow in its abilities to meet needs.

- Train new team members on the solution.

- Help keep the team focused on the 80 percent to 85 percent of met expectations rather than the 15 percent to 20 percent of missed expectations.

There is a cost for this strategy, but it is well worth it.

- *Productivity Software.* Most of the major software providers, like Microsoft and Adobe, provide learning libraries with complete instructional videos for using their software. These resources often are included with your software license. You may also find resources by searching online video libraries like YouTube. Accountability is wise, so we recommend setting up a requirement (minimum number of sessions or even specific courses) that must be completed within a specific time period for each team member.

A Great Example

One church's staff productivity and efficiency impressed us more than any other. A staff member explained what the church does, and we now recommend this approach often:

- *Identify your specialists.* Look for team members who have a real passion and proficiency in each of your software solutions (word processing, spreadsheets, databases, and so on). Make these your specialists! Make certain you don't choose the same person for more than one or two solutions; spread the responsibility and acknowledgement around. Doing so also helps keep you from losing your entire knowledge base if that one person leaves.

- *Invest in your specialists.* When you identify your specialists, tell each one they have a budget for training to get even better with their respective solution. Add the requirement of outside training to their annual review to be certain they do it. This will also help them get training material ideas for training the rest of your team!

- *Schedule regular weekly, bi-weekly, or monthly training opportunities.* Here is how to do this well:

    - Require all non-credentialed administrative staff to attend the training.

    - Open the training sessions to credentialed administrative staff as well as non-administrative staff. Advertise among the team the topic of the next session so a credentialed or non-administrative team member can attend if it is something he or she wants to learn.

    - Rotate topics every session. For example, one week or month the topic might be how to do an email or text mail-merge, and another week or month the topic might be how to create a meaningful graphic dashboard using data from the ChMS database and accounting system.

This very powerful strategy costs little and accomplishes so much! The only way it works, though, is to have the full buy-in of the senior pastor or the person responsible for staff performance and development. Otherwise, some who need it most will always be too busy to attend, which will affect all others on staff, too. In addition, it requires a champion to coordinate the training and to see the plan through.

In addition to raising everyone's proficiency, other benefits of this strategy are that it will:

- Encourage people identified as specialists. Most people serving in these capacities are in behind-the-scenes roles and rarely get recognition for the quality of the work they do in fulfilling the vision of your church.

- Relieve IT staff from needing to train team members on how to use all of the solutions at your church. The specialist can become the primary person to help others learn how to use the solutions because the entire team begins to realize questions about a specific application are more appropriate for the specialist.

- Increase the ability of your entire team to fulfill the mission of your church.

A Security Training Tool

We ran across a security tool that trains your team on how to manage and handle email they receive.

Here's the challenge:

- There are many team members who feel they are too busy or can't be bothered with learning secure email practices.

- Many feel they know what they're doing already anyway.

- Many of those same people are the very ones who click on links in email that infect their computer or, worse, the network drive. Some get their identity stolen or (even worse from an organizational perspective), transfer tens of thousands of dollars because someone duped them into believing the pastor or ministry leader wanted—or needed—them to.

Welcome to the rescue, KnowBe4.com[24]! This service, like some others, lets you set up an account and design campaigns in which your team receives emails that look real, but are actually modeled after current spam. Anyone who responds inappropriately (clicks a link, completes an embedded form, and so on) gets put into a kind of limbo escaped only by watching a short, well-done online video explaining what he or she did wrong.

KnowBe4 is "best of breed," and we recommend subscribing to the Platinum Tier. KnowBe4 offers churches and ministries up to a 50-percent discount depending on the service level selected. Other services similar to KnowBe4 exist. Whatever your church chooses, be sure to ask about nonprofit pricing.

---

[24] KnowBe4, Inc., http://knowbe4.com. (last visited Feb. 1, 2024).

CHAPTER 14

How to Keep IT Staff in the Local Church

> The way a team plays as a whole determines its success. You may have the greatest bunch of individual stars in the world, but if they don't play together, the club won't be worth a dime.
>
> **Babe Ruth** [25]

The largest budget expense for most churches and ministries is the salaries and benefits paid to personnel. Sometimes another category, such as facilities, runs higher, but there's no getting around the fact that salaries and benefits consume a large—if not the largest—portion of a budget.

We are called to be good stewards, or managers, of the resources we've been provided, so this category of expense is full of "good stewardship" opportunities!

One of the most challenging aspects of budget stewardship involves attracting and retaining quality staff members over time—especially

---

[25] Babe Ruth, https://www.brainyquote.com/authors/babe-ruth-quotes (last visited May 7, 2024.).

those in positions that focus on what would typically be considered non-ministry skillsets. IT is one such skillset.

Typical Approach

When hiring to fill open positions, we rightly look for individuals who are motivated by our missional focus. We want people who share our beliefs—who are sold out to our mission, and who are well-qualified to do the work for which they're being hired even as they grow in their own spiritual faith.

We also rightly look for individuals we can afford, again driven by the charge to be good stewards. Some churches consult salary guides, but honestly, not enough do. Among those churches that do consult salary guides, many probably only focus on those published specifically for Christian churches and ministries.

That can initially work. When someone pursues a new job with a church IT Department, he or she mainly focuses on landing the position and those sources of salary information may be sufficient.

Over time, though, as this individual transitions from "new hire" status to a key, reliable member of the team, his or her skills and experience will grow. How can the IT Department ensure it keeps this valuable worker relatively satisfied, as far as compensation and benefits are concerned, given the "going rate" for this person is likely much higher outside of ministry? This is a crucial question because the loss of that team member would be difficult on the entire staff and ministry.

Our roles in the church and ministry world give us the chance to talk with a lot of IT people in both secular and ministry settings. Those who work for churches and ministries share these complaints most often:

- The work schedule (they have to work weekends when their friends are off and having fun).

- The high-stress demands by management to accomplish high reliability and to meet new needs—without a realistic amount of time or an appropriate expense budget for good equipment to do so.

- The low pay.

It's great to be motivated by the mission. But church and ministry leaders must understand the need to create effective supporting strategies that retain talented IT staff members—ones that sustain them well in the midst of ministry demands, especially as friends and peers in related positions outside ministry often enjoy higher pay and better perks.

Work Environment

Pastors and those in ministry have many friends and colleagues in similar roles. That's true for most of us: we often develop relationships with others in similar roles and interests.

When that happens in non-ministry disciplines such as IT, it creates a sometimes-challenging tension. That's because secular and ministry organizations approach IT so differently. Organizations that employ IT professionals vary widely in their approaches to work schedules, equipment budgets, professional development opportunities, and so on. Churches frankly don't tend to invest well in these areas for their teams. Here are a few strategies to consider that can help:

- *Work Schedule.* Evaluate the church's culture and look for opportunities to encourage non-ministry staff members to take weekends off. A couple of examples in IT would be the use of a larger team that could rotate weekends and/or use volunteers that can support most needs during weekend services. These would be people who can reconnect devices to the network that have gone offline, add paper to label printers used for children's ministry check-ins, and so on. They don't need to have network expertise, admin network passwords, or the keys to the kingdom. They just need to help system users keep going in the rush of worship services (see also Chapter 12, under the section titled *The Right Place for IT Volunteers* beginning on page 101).

- *Equipment Budgets*. Churches and ministries look for ways to minimize overhead so they can focus as much budget as possible on programs. That's appropriate. But church leaders would be wise to listen to IT team members when they give a cost estimate for something the ministry wants to accomplish. Rather than respond, "That's too high," leaders should consider the sophistication—and thus the higher costs—involved with technology needs. The technology for church projects and services is akin to configuring services for a convention center. Leadership, though, tends to assume the technological needs and tasks aren't challenging or won't cost too much, based on comparable, inexpensive set-ups for, say, a home. But this is not a home environment. The strategies, setups, and costs are very different.

- *Professional Development*. IT is a profession with constantly changing methodologies. Churches should fund professional development for IT staff members, even insisting they attend one or two annual conferences. They need to do this if they are going to continue growing in their skills and get recharged. One of the better opportunities for this in the church IT field is the Church IT Network.[26] This organization offers a high-quality annual conference every fall. And it is very inexpensive.

Compensation

Matthew Branaugh, attorney and editor of ChurchLawAndTax.com, once said, "One of the most effective tools available to churches is a salary survey." Enter ChurchLawAndTax.com's sister site, ChurchSalary.com, which offers, which offers helpful information on a variety of church positions, though it doesn't yet include non-ministry-discipline positions like IT.

For non-ministry disciplines like IT, it is important to research beyond salary surveys specific to churches and ministries. To get the best picture of what these positions are worth, research secular salary surveys to ensure whatever salary range you set is reasonable. Figure 5 below provides is a comparison of three 2023 salary surveys—two that are church-specific and one that is secular—for two typical church IT positions.[27] [28] [29]

	Robert Half	**Salary.com**	**LifeWay**
Help Desk Tier 1	$38k - $51k	$42k - $53k	(Office Personnel: Full-Time only searchable option) $41k - $48k
"Systems Engineer/ Network Administrator"	$93k - $142k	$80k - $168k	

Figure 5: Salary Survey Comparison

---

[26] Church IT Network, http://ChurchITNetwork.com (last visited Feb. 2, 2024).
[27] Robert Half, Robert Half 2021
    https://www.roberthalf.com/us/en/insights/salary-guide (last visited Nov 24, 2023).
[28] Salary.com, http://salary.com (last visited Nov 26, 2023).
[29] LifeWay, http://compstudy.lifeway.com (last visited Nov 26, 2023).

Please note the following observations:

- MinistryPay's and LifeWay's terrific surveys, like many in the church and ministry field, don't differentiate between highly skilled non-ministry disciplines like IT and all other non-ministry disciplines. Their surveys would be helpful in setting a salary for the "Help Desk, Tier 1" role, but are not much help beyond that role.

- Robert Half's surveys always feel a little high to those of us in ministry, but they more accurately represent what the compensation of these two common church IT roles would be in the open marketplace and they get us closer to the right salary ranges. This information especially demonstrates why so many churches have a difficult time keeping highly skilled team members, such as a network administrator, from looking elsewhere for income and employment.

- Starting salaries vary from city to city and region to region. According to Robert Half, the reasons are cost of living, scarcity of top talent, and more. And the variances can be extreme! In some cities like San Francisco and New York, churches and ministries should add more than 40 percent to these numbers! And churches and ministries in places like Kalamazoo, Michigan, and Stockton, California, should reduce those numbers by about 20 percent. It is important any survey figures are adjusted accordingly for geographic location.

- This is not to say that a church needs to match the compensation levels available with secular employers. Some will say that the ministry component (the IT person's buy-in to the mission of the organization) should override the IT person's drive for higher compensation. Mission should be a strong component—especially in an individual's early years of church employment.

- But if the IT person has proven to be invaluable, then the disparity between the church and secular salary guides must be balanced—that is, if the church wants to keep the individual on staff long-term.

- Remember, the cost of replacing a key team member is higher than the cost of keeping one by increasing his or her salary and/or benefits. Good stewardship can include reasonably increasing those costs to keep a quality IT team member.

CHAPTER 15

IT Staff: Insource or Outsource?

> Only do what only you can do.
>
> **Andy Stanley** [30]

At one time, there might have been one computer in a church, most likely in the accounting office. The use of computers has grown dramatically since those early days of church computing. Now every staff member has at least one computer—or access to at least one.

With the expansion of computer use, each church will likely reach the point of realizing regular computer help is needed. That is when the church looks at bringing in a specialist.

But should the church "insource" by hiring the specialist as a full-time staff member? Or should the church "outsource" by hiring a firm or someone as an independent contractor? Understanding the five different IT roles in churches and how they differ (see Chapter 1) is helpful when deciding whether to insource or outsource.

Outsourcing some of the support responsibility usually is a better strategy!

---

[30] Andy Stanley, *Next Generation Leader: Five Essentials for those who will shape the Future* (2006).

What IT Roles Are In Play?

In multisite megachurches, it makes sense to hire all five IT roles as in-house specialists.

What about small- to large-sized churches (under 2,000 in average weekly worship attendance)? The answer changes to *maybe*. The decision revolves around what IT disciplines are the most heavily used. Those disciplines are (see Chapter 1 for more details):

- Web and graphic design;
- Audio/video;
- Social media;
- Data infrastructure; and,
- IT help desk.

In small- to large-sized churches, one person usually fills these needs, even though it is unlikely the person has the appropriate depth in all five disciplines. That's when it makes great sense to outsource some of the responsibility. Doing so allows the church to benefit from large strategies,[31] even though the church cannot afford to hire someone full time who possesses the levels of expertise needed to bring those large strategies in.

Acknowledging Our Bias

We need to pause and acknowledge our bias here. We each sensed the Lord's call to help Christian churches and ministries in this area, and established MBS, an outsource resource. We believe MBS helps fill

---

[31] By *large strategies*, we mean those IT strategies developed through working with many organizations to improve system reliability, reduce overall costs, and minimize the distractions problematic systems force on staff. The more systems one has engineered and had to support over time, the larger the strategy becomes to deliver reliability.

the data infrastructure side of outsourcing well (and by God's grace, its clients seem to agree). So, we are believers in the outsourcing model.

We have had the privilege of serving and consulting countless churches and ministries nationally and internationally for more than three decades. We have seen patterns of how churches approach this issue. Some of those approaches work well. Some do not. Though this is not a commercial, given that perspective, we offer a few thoughts here that may help.

Strategies that Increase IT Staff Size

MBS consulted with a large ministry that wanted to know if its IT Department was what it should be. The ministry wanted us to evaluate its department's efficiency, size, and quality. We conducted what we call an IT Audit to accomplish that. What we discovered was surprising:

- The entire ministry staff was about 100 people. Sixteen of them were in the IT Department! The IT staff salaries and benefits cost the organization more than $900,000[32] annually!

- The ministry had developed a few IT philosophies, or strategies, that drove the department to grow to such a large size:

    - Rather than buy enterprise-class hardware of an appropriate quality, the department bought inexpensive parts and built the computers and servers to save money. The true cost to maintain that hardware, though, included several full-time salaries—year after year.

---

[32] The study was in 2000, and the amount was more than $500,000 annually—that would be about $900,000 at the beginning of 2024!

- The ministry's corporate culture included inviting the entire staff to dream about what IT tools they'd like that would help them in their ministry. More than half of the IT Department consisted of programmers to create the apps the rest of the staff requested. The programmers also had to maintain those numerous apps. Surprisingly, no one checked to determine which were still in use!

There are strategies that can reduce the need for IT staff, and can save a lot of money. Here are a few:

- Buy enterprise-class hardware and get at least a three-year warranty from the manufacturer with any of that hardware. Though letting desktop and notebook warranties expire after three years makes sense (since they are lower-cost items), servers—which are computers the entire ministry must be able to depend on 24-7-365—should always be kept under warranty. Dell, for example, will extend the warranty on a server to a total of seven years! After that, a server should be retired or put in a non-critical role.

- Eliminate custom apps whenever possible. They are costly to create and even costlier to maintain.

- Outsource all but what you can't. Though we encourage you to keep Tier 1 help-desk functionality in-house, other tasks, including engineering, programming, and pulling cable, and so on, are good candidates for outsourcing.

When to Outsource

Though it depends on how much small-to-large-sized churches use graphics in their communications or A/V in their worship services, it seems the people most often sought for insourcing have great strengths in these disciplines. That most likely happens because, oftentimes communications and worship services are the first areas with felt needs for IT expertise. While these people may use vendors or church members to augment their skills in graphic artistry, for some reason they are often reluctant to do so in the data structure side of IT.

The assumption is typically that the IT needs for data structure are simple: give everyone full access to whatever they want and let the staff mostly manage themselves! These systems are often approached with a BYOD (Bring Your Own Device) strategy.

That is okay, unless there are also servers, public WiFi, and data responsibilities to manage; if so, it makes sense to outsource. While a church may be able to get by with outsourcing as needed for web and graphic design, A/V, and social media needs, a much more formal approach is necessary for outsourcing data infrastructure needs.

Who Can You Trust?

In Chapter 12 we described the best approach to using volunteers in IT, which is to address the three facets of church IT needs: sophistication and complexity, mission-critical dependence, and budget sensitivity (see "Church IT Complexity" beginning on page 121). With those facets in mind, a good outsource resource in the data infrastructure discipline should meet the following minimal criteria:

- *Data Infrastructure Expertise.* Because a church's IT needs are sophisticated and complex, a good IT outsource resource should have deep experience in strategizing, designing, and supporting local area networks and cloud-hosted solutions for many organizations (not just a few, but many dozens) of a size similar to your church and larger. This large perspective requires strategies that minimize downtime and support needs. If your team needs to use your church's systems without distraction, this is important.

- *Church IT Expertise.* Because a church's IT needs are mission critical and budget sensitive, a good outsource resource should have the experience of working with many churches and ministries. Otherwise, it is likely the firm will not understand the mission-critical nature of how church teams rely on their systems to fulfill their calling, while also meeting the constant flow of weekly deadlines. The natural inclination will be to utilize costly over-spec'd systems the firm sells (which wastes funds) or consumer-class systems that cannot meet the reliability needs of the staff. Neither is a good use of funds, and those with church IT expertise know that.

- *Church Software Expertise.* Just like hospitals use software unique to their needs, as do accounting, legal, and engineering firms, churches use unique software to meet their needs. Whether it's your ChMS database, a nonprofit accounting package, presentation software, or event scheduling and registration software, the outsource resource needs to know how those systems work and what they require. Some have very specific and unique requirements, and delivering reliability means knowing how to configure the systems so the underlying infrastructure is supportive to them.

Screening IT Vendors

Most churches around the US will look for *local* IT vendors to save the funds that would have been spent on travel with an out-of-the-area vendor. In the previous section we gave you four things to look for in a vendor, but what additional factors should one look for? Here are a few more thoughts that may help:

- Ask for references of organizations similar to yours (number of servers, WiFi access points, desktops, and notebooks/tablets).

- Call a few of the references and ask:
    - How long they've been using the vendor.
    - If they required a long-term contract (avoid any that require anything longer than a few months).
    - If they provide weekend support for no extra charge.
    - If they had to look for a new vendor, what would they do differently?

- Check any customer satisfaction resources that might be available. One you might check with is the Better Business Bureau where the vendor is based.

One key consideration when looking for an IT vendor is to look for one with multiple staff. We recently spoke with a church that used an IT vendor that was a one-person shop. The church felt he served well, but he did not document the church's system. When he died, it was up to the church, or the church's next IT vendor, to figure out how the system was configured. That took time and cost money, and until it was completed, the church's team suffered from more costly support due to the extra time needed for discovering the details of each support-related need.

We previously mentioned that MBS is an outsource resource. It's important to note that there are other firms who also focus on serving churches well, regionally and/or nationally. Here is a list of five, in alphabetical order (MBS included, of course!):

- Acts Group (http://actsgroup.net)
- BEMA (http://bemaservices.com)

- Enable Resource Group (formerly Solerant) (https://enableresourcegroup.com)
- Higher Ground (https://www.acstechnologies.com/higher-ground/)
- MBS, Inc. (http://mbsinc.com)

Remember that the homework your church does upfront with IT outsourcing can make a crucial difference with your church's short- and long-term IT success. Research the data infrastructure, hardware, and software expertise the vendor possesses with churches. Talk with other clients (ideally other churches and ministries) to find out the vendor's strengths and weaknesses. And make sure the vendor employs multiple people to ensure continuity and consistency.

CHAPTER 16

Who Owns Your Public DNS Record?

> Dovorey no provorey. That means trust, but verify.
>
> **Ronald Reagan** [33]

DNS stands for Domain Name System, and it is the worldwide method to help navigate the internet. For instance, opening a browser and typing in the address field "mbsinc.com" will take you to MBS's website.

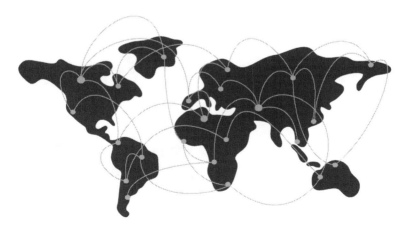

Figure 6: DNS Servers: A Worldwide Network

---

[33] President Ronald Reagan, "Remarks at a Senate Campaign Rally for Christopher S. Bond in Springfield, Missouri," (Oct. 23, 1986).

DNS is how the internet knows the location of MBS's website. It stores the server Internet Protocol (IP) address for each website and more.

DNS runs on countless servers around the world so that, regardless of where your website is hosted, someone anywhere in the world can connect to it.

Your church's public DNS record is the record the internet uses to point others to your website, your email server, and potentially much more. So, who owns your public DNS record and why does it matter?

Why is DNS Record Ownership Important?

One way to think of DNS is to think of your street address. It is important in how people in your community reach you, and if your address suddenly disappeared from all databases, making it impossible for someone to find you, that would be a problem. In today's culture, if an attempt to find you simply takes too long, an individual will likely lose interest and move on.

That is exactly what happens when someone tries to get to your website or send you an email. With even a minimal delay, let alone a problem, people will simply move on.

Having control of your public DNS record means you can fix a problem if one occurs, and you can keep one from developing in the event of an employee or member disagreement or a controversy in the church.

Public DNS Record Access

Technology is constantly changing. When you change or add IT services (like email or instant messaging), change internet service providers, change server configurations, and so on, doing so often requires making a change to your public DNS record. More likely than not, there will come a time when someone working with your website or email system will want to make a change to your public DNS record. We often encounter this as we begin working with new clients. Many times, no one at the church knows how to access and modify the DNS record. The most common reasons why include:

- A volunteer or staff member created the church's first website or email server and set up a DNS record to make it reachable. At the time, it was done under this person's name with his or her credentials (user ID and password), and he or she never thought to properly document it or transfer it into the church's name.

- When the church set up its first website or email server, the web host set up the DNS for it.

Both situations need to be addressed, and here is why:

- In the case of DNS record ownership by a volunteer or staff member, what happens when the relationship with that person and your church or ministry ends? For various reasons, volunteers sometimes change churches, as do team members. In either case, if DNS ownership moves with the individual, making future changes will be more difficult than it should be. It could be even more difficult if the person has died, and the record ownership is in probate or with heirs who are not friendly to your church.

- All businesses—churches included—need the freedom to change vendors at any time and for any reason. If a web host holds ownership to the church's DNS record, making a change in that relationship can be problematic. An employee of the web host could try to use the DNS record as a bargaining chip to keep your church's business there.

What Is The Best Approach?

MBS's new clients often ask whether MBS wants to take on their DNS records ownership. MBS's answer is always no. It needs access to it, but the church should maintain ownership of its DNS record.

We usually recommend the church move its DNS record to DNS Made Easy.[34] Its annual fee is reasonable, and this choice keeps the DNS record in a neutral environment. The church can grant or revoke access to whomever it wants as needed.

If you don't know who owns your church's DNS record, this should be a high priority. Find out where it is, who has access to it, and take ownership of it.

---

[34] DNS Made Easy, http://dnsmadeeasy.com. (last visited Feb. 5, 2024).

CHAPTER 17

Disaster Recovery and Business Continuity

Be prepared!

Boy Scout Motto [35]

Disasters of all types happen, and they can happen at any time. That's why there's a robust news reporting industry—every day there are new disasters to talk about! Some are natural, some are manmade, but they all disrupt lives.

Figure 7: First Baptist Church in Gulfport, Miss., after Hurricane Katrina hit

September 11, 2001, changed many things in the United States. More lives were lost in the attacks that day than were lost in the Pearl Harbor attack on December 7, 1941.[36] No one knows how many businesses also died that day, but it's safe to say that the

---

[35] Boy Scouts of America, http://usscouts.org/advance/boyscout/bsoathlaw.asp.
(last visited 11/2/2023).

[36] National Commission on Terrorist Attacks Upon the United States, The 9/11 Commission Report Executive Summary 2, (July 22, 2004).

number was large because the data necessary to sustain their operations (accounts receivable, accounts payable, various documents and spreadsheets, databases, and so on) was not properly backed up.

The same is true of major *natural* disasters, such as hurricanes, tsunamis, tornados, earthquakes, and fires. The photo on the preceding page shows a church that was severely impacted by Hurricane Katrina, reinforcing just how disruptive disasters can be to church operations.[37]

Many appropriately consider disaster recovery to be one of IT's highest priorities.

Backup Strategies

The goal of any good backup strategy is to protect data from threats that could destroy it. There are many backup strategies. The best are automatic, comprehensive, regularly tested, and include an off-site storage component. Here are details on those:

- *Automatic.* Manually initiated backups are guaranteed to fail. Backups aren't needed very often, and any computer user can easily fall into the *I don't have time today* syndrome, which then develops into a backup system that is never run. Backups should run regularly, automatically, and unattended. There are software solutions that can automate backups. Our favorite is Veeam.[38]

- *Comprehensive.* We are not fans of partial backups, sometimes referred to as incremental backups. The problem is that when restoring data, time is usually tight. The longer it takes to find all the needed files and restore them, the more stressful the sit-

---

[37] First Baptist Church in Gulfport, Miss. (2005). Photo by Greg Warner. Used with permission.
[38] Veeam Software, http://veeam.com. (last visited Feb. 5, 2024).

uation becomes. A comprehensive backup avoids that stress. *Regularly Tested.* There's nothing worse than running your first backup test while trying to recover from a disaster, and then learning that the backups weren't doing what you hoped.

We learned this the hard way when we helped a church replace the Network Operating System (NOS) of a server. Our first step was to run a comprehensive backup and have the backup software confirm that the backup was good by comparing file names, dates, and sizes. Once that was done, we deleted the server's partitions (effectively erasing any possible data on the hard drives), then we installed the new NOS, installed the backup software, and finally restored the backup we had just created. We were shocked to learn that the backup was worthless, even though it passed the compare test! All the characters in the files were null characters (null characters look like this. ☐ They're empty)! Using some tech tools, we restored most of the data, but it made for a long and embarrassing night.

- *Include an off-site component.* A copy of a recent backup needs to be taken off-site in case there's a disaster large enough to take out the server room. The September 11 attacks on the Twin Towers and the flooding from Hurricane Katrina damaged or destroyed many on-site server rooms, underscoring the importance of off-site locations.

The off-site backup must be comprehensive. It cannot be an incremental backup. Some prefer online backup solutions, such as Carbonite or CrashPlan, to accomplish this. More about that shortly.

Backups come in many flavors and strategies. Differences between backups include incremental *versus* comprehensive, system level *versus* file-level, server-only *versus* server plus computers *versus* computers-only, and more. The server component is further complicated if your servers are virtual machines running on a host using a hypervisor like VMware or HyperV.

What We Recommend and Why

Networks and computers store data. Some data is mission-critical, but not all of it. Even still, a loss of any data can take a lot of time to re-create and would slow down productivity while doing so.

We recommend backing up the entire system. The objection sometimes voiced is that doing so would require a lot of capacity. Appropriate capacity is reasonably inexpensive, and it needs to be seen in a similar light as an insurance policy, which you hope to never use.

For networks with servers and computers, we recommend the following:

- Configure all computers so they save their data to the server by default. It will save time and money not having to back up each computer on the network.

- Back up servers that store data every night. Do not run incremental backups—only run full backups. If the servers are virtual, use a backup solution (such as Veeam) that backs up the servers so they can be quickly restored in their entirety if necessary.

- Back up to tape drives. A good strategy is to have at least 20 tapes (four weeks' worth), and to rotate them daily. Label them as Week 1 – Monday, Week 1 – Tuesday, and so on through Friday; Week 2 – Monday, Week 2 – Tuesday, and so on through Friday; and Week 3 and Week 4 similarly. This will be helpful, for example, if someone discovers that an important file is corrupt, and the last time they had a good copy of it was within the past three weeks.

Larger churches might choose to back up to an enterprise-class Network Attached Storage (NAS) or Storage Area Network (SAN) device. Those who don't have those devices should avoid backing up to external portable hard drives. External portable hard drives have many moving parts and are susceptible to breaking. This would render the backup on that device unusable without an expensive restoration process that may take more than a week and may work—but may not.

Backing up to tape is still Corporate America's preference. Even the large media studios still back up their very valuable media files, like movies, to tape!

- Take one tape off-site each week, rotating it with the tape from the previous week. If, like most churches, your biggest data processing day of the week is Monday or Tuesday, the backup made the night of your biggest data processing day is the one that should be rotated off-site each week. That way you can always get back to within a week of your data if a large catastrophe hits your church.

- If you're a large church and backing up to an enterprise-class NAS or SAN, you will want to set up a second NAS or SAN off site with data replication (an automated process in which one of these devices automatically replicates its data over a cable or internet connection) to accomplish the geographic separation component of your backup strategy. To save money, some churches partner with another church in which they replicate to each other's sites. For instance, Church A might put a NAS or SAN in Church B's server room, and vice versa.

- Test your backups at least monthly. Keep an accountability spreadsheet to identify the backup data that was tested, the date it was tested, what portion of the data structure was tested, and who tested it. The spreadsheet will help ensure that different sections of the data are tested and that, in fact, the testing happens.

What about Online Backups?

Online backup solutions are good for consumers' computers but not necessarily good for servers. The amount of data that would need to be transmitted over the internet to restore a server is usually so large that it could take days or weeks to download it all from an online service. Online services offer to send a copy on a hard drive if needed quickly, but the organizations we know that tried this approach were not pleased with the results.

Online backups are usually good for consumers who want to back up data files, such as photos. Some churches we know still use an online backup service to automate the off-site requirement and only use them to restore single files; they also have an on-site backup in case of a more severe disaster.

Setting an Appropriate Backup Budget

All technology costs something and thus requires a budget. Many who work in IT assume that all data needs to be quickly restorable at all times. That strategy, however, may be too costly and beyond the reach of most churches.

The best way to set the backup strategy and budget is to ask church leadership to prioritize various categories of data and assign a maximum timeframe within which data must be restored following a disaster. In addition to helping set an appropriate backup system budget, this will help leadership think in terms of business continuity. (Business continuity plans detail how to continue operations during and following a major disaster.)

Typical categories of data include audio and video files, databases, email, and productivity documents like word processing files, spreadsheets, and so on. You may even want to break those categories down further to help leadership with this task. For instance, word processing files might include letters, bulletins, policies and procedures, and so on.

Often, leadership will assign the highest priority to communication systems and databases—email, financial databases, and non-financial databases. An example would be that email, telephones, and ChMS databases need to be restored within two hours, and everything else within three days.

After leadership assigns time frames for data restoration, design a backup strategy to meet the requirements and present it for approval. If leaders think the budget is too high, ask them to adjust some of the time frames they assigned to the data categories so you can redesign the strategy accordingly.

Business continuity is leadership's responsibility and disaster recovery is IT's responsibility. The only way a backup strategy budget can be appropriately set is with leadership's guidance.

CHAPTER 18

The Twelve Months of IT

A small behavioral change can also lead to embracing a wider checklist of healthier choices.

Chuck Norris [39]

If disaster recovery is one of IT's greatest responsibilities, how does one stay on top of it? The answer for us is checklists. *Checklists* mean we don't have to try to remember to do certain things.

Setting up checklists can help eliminate the need to remember lists of tasks. Some might dismiss the need for checklists, but they are the key to avoiding letting things slip through the cracks. Our favorite tool for checklists is Microsoft Outlook, where we can create recurring tasks that show up monthly, quarterly, annually, and so on.

Monthly Task Checklists
Here is a monthly IT checklist for consideration

January
☐ Test backup quality by restoring a folder in the data structure that is different from any recently tested. If your backups are done by a

---

[39] Chuck Norris, "Chuck Norris Simplifies Healthy Living," WND (June 24, 2016), https://www.wnd.com/2016/06/chuck-norris-simplifies-healthy-living/ (Last visited Feb. 6, 2024).

cloud host, confirm with the cloud host that the backups are good and have been tested. If necessary, test them yourself.

- ❏ January is the month of new beginnings, so seek out team members who have been there less than a year and ask them if there's anything their system could do to better serve them in their ministry. Then pray for their success in your organization's ministry.

February

- ❏ Test backup quality by restoring a folder in the data structure that is different from any recently tested. If your backups are done by a cloud host, confirm with the cloud host that the backups are good and have been tested. If necessary, test them yourself.

- ❏ February is the month of love, so schedule a task that is an expression of love toward your user base. One idea: schedule a team of volunteers to come in on a midweek evening to clean users' mice, keyboards, and monitor/display screens. Encourage the volunteers to pray for the person who uses the workstation as they clean it, possibly for protection from the enemy—for them and their family—and for their effectiveness in ministry.

March

- ❏ Test backup quality by restoring a folder in the data structure that is different from any recently tested. If your backups are done by a cloud host, confirm with the cloud host that the backups are good and have been tested. If necessary, test them yourself.

- ❏ March is the month of St. Patrick's Day, whose associated color is green. Do something good environmentally, like remove debris and organize the server room, and clean dust from the fans and insides of the servers.

- ❏ Pray for God's protection of the servers, switches, related equipment, and the data they contain.

April

- ❏ Test backup quality by restoring a folder in the data structure that is different from any recently tested. If your backups are done by a cloud host, confirm with the cloud host that the backups are good and have been tested. If necessary, test them yourself.

- ❏ April starts with April Fools' Day, so check your system for peripheral equipment foolishly added by well-meaning team members (routers and switches, printers, and so on). Schedule follow-up time to help the users who introduced those devices. Help them understand the need to involve IT to determine their needs and coordinate ways to meet them. End the time with prayer for them, their family's protection from the enemy, and for their effectiveness in ministry.

May

- ❏ Test backup quality by restoring a folder in the data structure that is different from any recently tested. If your backups are done by a cloud host, confirm with the cloud host that the backups are good and have been tested. If necessary, test them yourself.

- ❏ May is the month we observe Memorial Day in the United States. Evaluate the church's hardware and software and identify which ones need their end-of-service date scheduled. Begin planning for their replacements—or their elimination if no longer needed.

June

- ❏ Test backup quality by restoring a folder in the data structure that is different from any recently tested. If your backups are done by a

cloud host, confirm with the cloud host that the backups are good and have been tested. If necessary, test them yourself.

❑ June brings the first day of summer in the United States. Help those who will be going on vacation by making certain essential data others may need to access during their absence is located on a part of the church's network that does not require sharing private passwords. Ask them if they'd like you to review how to set email and voicemail out-of-office greetings, and then how to turn them off again. Encourage them to completely unplug digitally while away, and pray for their safety and refreshment while on vacation.

July

❑ Test backup quality by restoring a folder in the data structure that is different from any recently tested. If your backups are done by a cloud host, confirm with the cloud host that the backups are good and have been tested. If necessary, test them yourself.

❑ July is the month the United States celebrates its independence, which is a good reminder that team members often act independently with regard to data and solutions. Encourage your team to share with you the ways they have done so (without reprisals, of course), so you can help them understand how the system is already designed to meet their needs in ways they may not have known (or help you understand ways the system needs to meet their needs better). End by praying for them and their family's protection from the enemy and for their effectiveness in ministry.

August

❑ Test backup quality by restoring a folder in the data structure that is different from any recently tested. If your backups are done by a

cloud host, confirm with the cloud host that the backups are good and have been tested. If necessary, test them yourself.

❑ This is the month when many return from vacation. This is a good time to review and upgrade systems so they're ready for the team's use as the fall, winter, and spring programs kick in. It's also a good time to pray for your organization's success in ministry.

September

❑ Test backup quality by restoring a folder in the data structure that is different from any recently tested. If your backups are done by a cloud host, confirm with the cloud host that the backups are good and have been tested. If necessary, test them yourself.

❑ September in the United States includes the Labor Day holiday. This is a good time to schedule training and offer support appointments for team members.

October

❑ Test backup quality by restoring a folder in the data structure that is different from any recently tested. If your backups are done by a cloud host, confirm with the cloud host that the backups are good and have been tested. If necessary, test them yourself.

❑ October brings Halloween ghosts and goblins, which is a good reminder that the enemy wants to do anything possible to disrupt your church and its ministry reach. In addition to checking system protection points (anti-malware, anti-SPAM, and firewall solutions), this is a good time to remind the team of the enemy's intent to do the ministry harm and enlist their prayers for the protection of the church's systems and data.

November

❑ Test backup quality by restoring a folder in the data structure that is different from any recently tested. If your backups are done by a cloud host, confirm with the cloud host that the backups are good and have been tested. If necessary, test them yourself.

❑ November includes the Thanksgiving holiday in the United States. This is a good time to thank supervisors for the opportunity to serve on the team. It's also a good time to express gratitude to our biggest support challenges for their efforts and impact on the ministry.

December

❑ Test backup quality by restoring a folder in the data structure that is different from any recently tested. If your backups are done by a cloud host, confirm with the cloud host that the backups are good and have been tested. If necessary, test them yourself.

❑ In December we remember God's greatest gift to us: forgiveness through Jesus Christ. This is a great time to think through each member of the team and pray for them—for them and their family's protection from the enemy and for their effectiveness in ministry—and then send them an email or text letting them know you prayed for them and are excited about the ways they contribute to the ministry.

Customize Your Checklist

You will likely have some additional items to add to your monthly checklists. Make certain to keep the list from being only tactical; keep spiritual components in it. You'll find they help you!

- Your team members will see you as a ministry partner rather than an adversary; and

- Doing so will help you avoid ministry burnout.

"No institution can possibly survive if it needs geniuses or supermen to manage it. It must be organized in such a way as to be able to get along under a leadership composed of average human beings," said Peter F. Drucker.[40]

---

[40] Peter F. Drucker, https://www.brainyquote.com/quotes/peter_drucker_130664. (last visited Feb. 6, 2024).

CHAPTER 19

The Security Sweet Spot

> Find the "sweet spot" between absolute transparency and non-transparency—it's called appropriate transparency—where trust is maximized with minimal disruption or risk to the ministry.
>
> **Dan Busby** [41]

Chapter 2 explained the importance of team members being thought of as customers of the church's IT Department. The chapter also described how that approach affects managing technology in churches. One topic in that chapter was giving *local admin* status to all team members (see "Here Are a Couple of Practical Examples" beginning on page 29), which some argue weakens the church's system security. On page 31, we mention two ways to overcome those security concerns.

Additional "customer satisfaction" issues for team members include your approach to password strategies, off-site file access, and WiFi access. How an IT Department approaches security affects each of these.

---

[41] Dan Busby, *Trust* (2015).

Security Is Important!

Strong system security is critical. Much of the data we have in churches is sensitive: the most common sensitive sets of data include counseling notes, board minutes, HR documentation, contributions information, and database records—including data on children. Churches need to do their due diligence to protect all of it.

Many people today question anything less than a full-access approach to church data. However, some of the data that churches possess can harm people—or the church—if let out into the wild.

Consider these scenarios:

- In some states, such as California, there is a state constitutional right to privacy. If a church employee gets terminated, and his or her termination details or contact information are not properly safeguarded by the church, the former employee could sue under state's law if he or she believed the lack of protection caused him or her harm.

- A church board works through a sticky issue, such as the possible church discipline of a member as detailed in Matthew 18 and references its discussion details in its minutes. That is sensitive information that could hurt the individual and the church if it were not adequately protected.

- A church's vendor realizes the value of the church's database and exports a copy of it. The vendor then rents it as a list in the community. Consider the contents of that database and the possible damage it could do! The database includes contact information, children's names, tithing records, life events, Social Security numbers of employees, and more. (This is a true story, by the way!)

Given the sensitive information involved, a church's data must be protected with firewalls and passwords.

The Right to Disappear from Your System

In 2018, the European Union's General Data Protection Regulation, or GDPR, went into effect. Many wondered how it would affect US churches and ministries. Those with a physical presence in an EU country are affected. Beyond that, while the EU doesn't have jurisdiction in the US, the law is written to operate beyond the EU's borders, so opinions vary on the exact legal impact GDPR has on churches or ministries.

But understanding why the GDPR was enacted may provide a more important insight. GDPR addresses whether someone can ask to be removed from a database and have a reasonable assurance they will be. People, for a variety of reasons, sometimes want to be forgotten or removed from a list. You could call this the "Golden Rule of Database Management."

Unless there is a legal reason why it would be unwise to remove someone from your database, why not simply accommodate this individual? Treating others like we'd like to be treated is always a win-win, and a great way to honor Christ in our data management.

With that in mind, note that you should consult with your church's legal counsel if you receive a request from someone who wants to be removed from your church's database. Some reasons why your church may not be able to remove someone include:

- This person made contributions that were posted to your system in recent years (some say the guideline is seven years).

- This person was an employee and has human resource and payroll records your church must keep.

- This person was a vendor who received payments from your church.

- This person was a volunteer who served in a position that required a background check.

Before you make a decision regarding a person's removal request, talk with your church's attorney to make certain the removal will not create unintended future liability for your church.

Password Strategies

Many churches have annual audits performed by CPA firms, often to satisfy the terms of building loans or to help demonstrate integrity in their finances to a watching world. Both are good reasons, and CPA audits are helpful in many ways.

The Sarbanes-Oxley Act of 2002 was enacted in response to accounting scandals at some very large companies. One of its many effects is that CPA auditors now ask about IT issues. This is part of their due diligence to ensure things are correctly handled. Most of those auditors, however, do not have professional IT training or experience, and simply work through a series of questions and record the responses.

Through audits CPAs have heightened everyone's sense of appropriate IT security. Some things they brought attention to are very good, such as locked server rooms. However, on password strategies, they may have hurt us by requiring password changes every ninety (90) days. The practice of changing passwords so often in churches actually *lowers* system security! Church team members are like most computer users. When they change their passwords they often write them on Post-It notes, or tape them to their monitors and displays, or place them under keyboards or a desk drawer.

In a March 2, 2016, post on the US Federal Trade Commission's website, policies requiring regular password changes proved "less beneficial than previously thought, and sometimes even counterproductive."[42] They go on in that post to reference two university studies that caused them to draw the conclusion that "frequent mandatory [password] expiration inconveniences and annoys users without as much security benefit as previously thought, and may even cause some users to behave less securely."

We have said that for years!

We recommend the following password policy:

- The password must be at least 14 characters long. Think in terms of phrases instead of words. A good example is a favorite verse or song phrase. Something like "inthebeginninggod" (without the quotes).

- The password must meet certain complexity requirements:

    - It should not contain the user's account name or parts of the user's full name that exceed two consecutive characters

---

[42] Lorrie Cranor, "Time to Rethink Mandatory Password Changes," Federal Trade Commission (March 2, 2016), http://ftc.gov/news-events/blogs/techftc/2016/03/time-rethink-mandatory-password-changes (last visited Feb. 6, 2024). See also "Surprising New Password Guidelines from NIST," Password Ping, http://passwordping.com/surprising-new-password-guidelines-nist/ (last visited Feb. 6, 2024)

- It should contain characters from **EACH** of the following four categories:
    - English uppercase characters (A - Z)
    - English lowercase characters (a - z)
    - Base 10 digits (0 - 9)
    - At least one non-alphabetic character
    (for example ! $ # %)

- Using our previous example, a strong password could become "1nthebeginningGod!" (without the quotes—and notice how the lower-case "i" became a numeral 1).

- Passwords should not use words or phrases that are part of any decorations around the user's computer. Someone with physical access to the office or home office could easily try phrases on posters, pictures, plaques, and so on.

- Do **NOT** re-use this password anywhere else. Passwords should be unique and never re-used for any other sites or services. This ensures a compromised password on one site or service affects only that site or service—and not more.

Passwords of this length and complexity are virtually impossible to hack. This means required password changes become unnecessary. They only become necessary when the user shares the password with someone else or the church becomes aware of a compromised password.

We recommend setting your systems so that a user gets five (5) login attempts that fail before the account becomes locked. This is for your protection in the event a hacker is attempting to guess a password. After five failed attempts the lock should last twelve (12) hours unless the user submits a help-desk ticket requesting an unlock. After twelve hours pass the account automatically unlocks and the user receives five more chances to login.

Multi-Factor Authentication

Multi-Factor Authentication, or MFA, as it is sometimes known, engages two or more kinds of authentication factors consisting of:

1. *Knowledge*: something only the user *knows* (like a password, passphrase, or security question);
2. *Possession*: something only the user *has* (an identity card or USB key); and
3. *Inherence*: something only the user *is* (biometrics like a fingerprint, finger veins, retina, or facial recognition).

Another form of MFA is Two-Factor Authentication (sometimes called 2FA). As the term suggests, it uses two of the factors mentioned above.

These systems can be set up in various ways. One common method is to use a 2FA process in which one of the authenticating factors is sent to another device and entered—such as a code sent to a smartphone.

This may be a good way to increase system security, and churches are wise to consider it because it can prevent some attacks. But most church and ministry system users don't like the additional steps. There are also technological reasons why the increased security may be more of a *perceived* security improvement than an actual improvement. For instance, many "account recovery" processes bypass mobile phone 2FA, providing a potential workaround for a hacker. That said, MFA is worth considering.

Firewalls

A firewall is a solution that keeps unwanted outsiders from accessing your computer systems. We need to guard against internet bots (programs) and people—they are the two major threats from the outside. Here are more details on both:

- Internet bots are little apps that roam the internet looking for exploitable ports into networks and computers. They are incessant, and their technology is constantly morphing into new threats. For that reason, it is essential that firewalls receive constant updates to help them identify new technology threats. Many firewall companies offer an annual subscription to meet this need.

- People who want to gain access into networks and computers include former employees, family members, and others in the community. Some have no malicious intent, but some do. The fact that some have bad intentions is why it is essential to have a policy of never sharing passwords.

Firewalls, aided by good password strategies, help protect churches from these threats. A good firewall and a good password policy help protect sensitive data from those outside the church who shouldn't have access to it. Firewalls and password policies can enforce access rules that augment the network security systems in place.

There are software firewalls and hardware firewalls. Either can be good, depending on their spec. Our favorite spec is SonicWALL firewall series. SonicWALLs provide the features that most churches need at a reasonable cost while not overcharging them for many features they will likely never use.

One of SonicWALL's features we like is that it can also filter internet content; this prevents people from using the church's WiFi to access websites the church considers inappropriate.

In Figure 8, note that the internet connection comes in to a modem (sometimes called a router). Rather than plugging computers directly into the modem or router, the next device in line is the firewall, which then connects to the switch where all the computers are connected. By doing this, nothing comes in from the internet or goes out to the internet without first going through the firewall.

Off-Site Access

System users today are very mobile and need access to data while working, studying, or prepping while away from the church. There's more than one way to provide secure off-site access. Here are a few of the most prevalent ways:

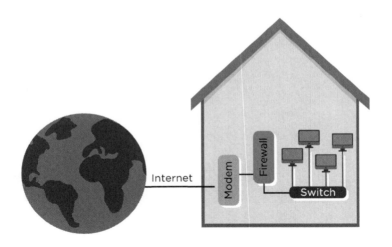

Figure 8: Diagram of a Firewall's Position

- *VPN Access.* A Virtual Private Network (VPN) is the highest level of security for off-site access used by churches. Although it connects over the internet, it is a private network. A VPN accomplishes a higher level of security by creating an encrypted link, and it requires the off-site system to run an app that has the digital key to unencrypt the link. Like all IT solutions, it has strengths and weaknesses:

    - *Strength*: VPNs provide the highest level of simple security for off-site access.

    - *Weakness*: VPNs require the off-site user to run an app to establish the VPN link, which is often more than what most in ministry can successfully cope with technologically. It's not that the app is difficult to run; it's that it requires an extra step to gain access that is often forgotten and generates frustration and support requests. VPNs are not intuitive.

- *Hosted databases.* Many ChMS databases are now hosted and available via a web browser (like Firefox, Safari, Edge, and so on). This automates the security when they are hosted on secure servers (an important requirement for a hosted ChMS is that its hosting servers are using security protocols). This, too, has strengths and weaknesses:

    - *Strength*: Data security access is automated, making it easily available to off-site staff and volunteers.

    - *Weakness*: The only data available is what is in the database. If there is additional data, like spreadsheets or documents, for example, they may not be available in this manner.

- *Remote Desktop Connection*. Data is available on a server that requires a connection, but the connection may not be highly secure. It looks like accessing data on a typical Windows server.

  - *Strength*: The interface is familiar and more intuitive than the process of establishing a VPN, and it looks like typical data access on a Windows server.

  - *Weakness*: Remote desktop sessions can be susceptible to what is called a "Man in the Middle Attack," in which someone who is monitoring area internet traffic might intercept data being transferred. Some IT people combat this by adding a remote gateway that shields the actual server from the public, adding a security certificate, and securing the connection to help strengthen the security of this method.

- *Remote Data Synchronization*. Many use an app, like Dropbox, Google Drive, OneDrive, or Owncloud, to synchronize data between devices. The net effect is that data can be stored on the mobile device *and* on the server, with changes made to it synchronized in real time if both are connected via network or internet connection.

  - *Strength:* This method, once it's configured, is fully automatic and requires no user intervention. On devices with storage capacity, like notebook computers, data can be stored locally on the mobile device and accessed as needed with or without an internet connection; changes will be synchronized automatically if or when next connected.

  - *Weakness:* Data on mobile devices increases risk if those devices are not adequately protected by their user, often getting lost or stolen. Thus, security training is recommended. Also, some of these apps do not have good security, so if they require storing data on a public cloud server, they may not be a good solution. Dropbox is one whose security is questionable. That is why we prefer Owncloud (private cloud server with more tightly controlled access) or OneDrive.

Mobile Device Vulnerability

By their very nature, mobile devices (smartphones, tablets, and notebook computers) are more vulnerable than other devices. They are more easily stolen or lost, so a strategy should be in place that allows a church to delete any sensitive data from a lost or stolen device the next time it connects to the internet. Some data synchronization services help, as do some mobile device operating systems. The strategy to delete sensitive data could be driven by the operating system and/or apps on the mobile device.

Churches should also have a defined process to delete sensitive data from an employee's mobile device upon resignation or termination.

Internet of Things (IoT)

Many now have internet-connected devices in their homes. We're not talking about computers, tablets, or smartphones—we're referring to digital assistants (like Amazon Echo, Google Home, and Apple HomePod) and home devices and appliances like thermostats, refrigerators, door locks, and more.

Upon purchase, most of these devices possess very limited security, and they're easy targets for internet programs (bots) that want to access and control them.

In October of 2016 a Distributed Denial of Service (DDoS) attack caused trouble for countless people worldwide. A DDoS attack occurs when web-connected servers, computers, or devices are hit with a huge number of false connection requests that essentially freeze out all legitimate requests.

In a forensic analysis after the DDoS attack, it was discovered that the attack was caused by a botnet (malware-controlled devices) made up of 100,000 devices! Further analysis found that it could have been much worse because there were more than 500,000 devices in the botnet, and only 100,000 were used.

What should you do with your IoT devices? Here are some steps you can take to protect them from being accessed and controlled:

1. Contact your Internet Service Provider (ISP) and have the ISP demonstrate to you that the modem or router it provides for connecting your church to the internet has security turned on and that it's set to the highest level (if there are multiple levels available).

2. Change the default password of the device.

3. Consider putting a firewall between your modem or router and the rest of your devices that connect to the internet.

CyberSecurity Risk Assessment

Churches also should consider hiring a firm that specializes in cybersecurity risk assessments to look at their systems and issue a report of strengths (what you're doing well) and weaknesses (opportunities for improvement). There's almost always *something* that can be improved, and having an expert set of eyes looking for those opportunities is helpful. You may want to check with your church or ministry's insurance carrier to see if it can recommend a cybersecurity firm to perform such an assessment. Doing so may even help reduce your church's insurance premiums!

CHAPTER 20

The Value of Standardization

Manage for results.

John Pearson [43]

When earning his business administration degree (with a focus on management), Nick formed a philosophy that shaped the way he approaches the use of computers: *Good managers get the best possible results from limited resources.*

Good Management

As Christians, we want to hear at the end of our earthly journeys, "Well done, good and faithful servant. ... Come and share your master's happiness!"[44] Our guess is that you feel the same way! So, how does a faithful servant, or in today's vernacular, a *good manager*, earn such praise?

The first important lesson is one taught by Jesus himself. He was a servant leader. The Last Supper is a great example—Jesus lowered himself to wash the feet of his friends. He even served the one he knew would betray him! Serving everyone with love and humility is essential.

---

[43] John Pearson, *Mastering the Management Buckets* (2008).
[44] Matthew 25:21

In every job, managers wrestle with limited resources. Whether it is restricted cash flow, teams that are too small for the task, equipment that is less than optimal, or something else, there is always at least one challenge. Yet, there is a job to do and goals to accomplish. Good managers, like race car drivers, do their best to keep RPMs and speeds as high as possible without blowing up the engine.

Similarly, a servant leader does his or her best to keep the team motivated and moving forward. Because servant leaders have an underlying posture of love and humility, they don't push their teams beyond what can be endured. They ensure team members are paid enough to care for themselves and their families, have enough downtime to get refreshed, and have appropriate tools available to help them with the fewest distractions possible. That last point is where technology standardization has its impact.

Standardization

As we began deploying larger and larger networks, we developed standards that helped us configure workstations. These standards helped us create workstations that fulfilled our clients' needs to accomplish their ministry goals without distractions. Over time, we found that some computer configuration settings helped and some hindered.

As we began to identify optimal configurations, we found it challenging to deliver those consistently on every workstation. It was easy to miss a setting or two on any given computer, and any lack of consistency resulted in *support tickets*. We saw the value in achieving standardization and needed a way to deliver it.[45]

---

[45] Figure 9 on page 161 is intentionally too small to easily read. The intent was to show how detailed our checklists are.

We developed a checklist system for each computer configuration: servers, desktops, and laptops/notebooks, and for each Operating System, like Windows and macOS. Having a method of delivering a standardized configuration, we noticed that:

Figure 9: MBS' Setup Checklist for Windows 10

- *Client satisfaction* continued to increase, keeping pace as we began working for larger churches and ministries.
- *Support tickets* remained low despite the larger number of systems we were supporting.

Focusing on high-quality standardization helps keep our clients' support costs low. As we think of their call to build the Kingdom, that translates into fewer distractions and higher results. It means we're helping them achieve the best possible results with limited resources!

MBS' Setup Checklist

Here are some excerpts from our checklists for Windows 10 and macOS that you may find to be a helpful start:

- Windows 10
    - Windows Explorer Configuration & Actions
        - ❏ At the Root of the C drive, go to the View tab
        - ❏ Set View to List (this is a better view for most users)

☐ Check File Name Extensions (most users want to see file extensions, and it helps with support)

- Control Panel Settings & Configuration
    - ☐ Sound: click the Sounds tab, then click Sound Scheme, then click No Sounds (helps to minimize office distractions)
    - ☐ Sound: while on the Sounds tab, uncheck Play Windows Startup Sound (helps to minimize office distractions)
    - ☐ Display: Adjust Resolution and/or Text Size
    - ☐ Screen Saver: set to none (helps reduce support issues)

- macOS
    - Finder Preferences
        - ☐ On the General tab, set New Finder windows to open to the user's home folder (speeds up finding files)
        - ☐ On the Sidebar Favorites tab, remove all checks except desktop and home folder (speeds up finding files and is less confusing)
        - ☐ With the Finder window open, click the View menu option (at top of the screen). Check Show as Columns, Show Path Bar, and Show Status Bar (columns is a better view for most users, and the Path Bar and Status Bar are very helpful in folder navigation and support)

    - Safari Preferences
        - ☐ Set downloads to automatically go to the Desktop (helps the user remember to delete downloads that are no longer needed, saving storage space)
        - ☐ Set the church's standard home page

We also include a section detailing which apps should be installed on every system (like Microsoft Office, SentinelOne, and so on).

This type of standardization helps users become more efficient, and it helps the support team when it responds to users' requests. The users can change whatever they want, of course, but at least they each have a common starting point and are more predictable as a result.

The mistake many make is not standardizing, which means every machine is different based on how it was set up by the manufacturer and may not be properly set up to run well for the user. Non-standardized systems tend to be more confusing to users and more challenging to support.

Mobile Device Management

Mobile Device Management (MDM) is software designed to accomplish standardization across multiple operating systems and on various types of hardware. It is a great concept! Imagine managing any device in your system—whether face-to-face or remotely—and delivering operating system and app updates fully configured! Many churches and corporations successfully use this technology.

CHAPTER 21

Changing Paradigms: The Cloud, BYOD, Social Media

<hr>

The price of doing the same old thing is far higher than the price of change.

Bill Clinton [46]

<hr>

The iPhone and the Android have been around since 2007. While there were smartphone-*like* devices (the Palm Treo, for example) prior to that, it wasn't until the iPhone and the Android that smartphones became mainstream. Those devices forever changed the way we think about using technology today.

Churches occasionally ask us to help map out a ten-year IT strategy. In response, we point to recent technological changes, such as smartphones, tablets, and wearable devices, as examples of how fast things change. Ten years is too far out to plan! We suggest a two-year plan (the plan is a list of things they need to accomplish), which is based on a three- to five-year horizon (identifying the general direction where they want their ministry to go, sort of like a compass heading.

---

[46] President William Clinton, *Public Papers of the Presidents of the United States, William J. Clinton, 1993 Book I: January 20 to July 31, 1993* (1994).

Four major paradigm changes are wildly affecting the planning for IT strategies and structures: the cloud, the "Bring Your Own Device" (BYOD) movement, social media, and artificial intelligence (AI). AI is now so significant, we address it in Chapter 22.

The Cloud

The cloud is a paradigm shift that is mature, and most churches are taking advantage of its benefits.

Most of us, when we think of the cloud, think of social media platforms, email solutions, and online databases or backups. Those are important and make a difference. But the greater potential impact of the cloud is more than those services, and that impact is what really helps churches the most.

For decades, MBS has installed and configured local area networks (LANs) for churches and ministries nationwide. Those LANs require major capital investments in hardware, software, and engineering, which significantly impact cash flow. The cloud can remove that large capital expense. This is one of the cloud's greatest potential improvements for churches because it allows them to focus more of their cash flow toward ministry programming.

How much capital expense does it remove? Compare the costs in Figure 10 on page 167 of two real situations for a large, single-site church (with about 2,000 in average weekly attendance). The ChMS provider already hosts the church's ChMS.

Item	Locally-Based	Cloud-Based
Hardware	$12,830 2 Servers (1 as a host & 1 for a bare metal backup), 2 UPSes, Backup Drive and Tapes	None
Software	$3,415 VMware vSphere Essentials, Windows 2022 with CALs, Exchange 2019 with 50 mailboxes, Veeam backup software	None
Engineering	$30,970	$6,720
Total Start	$47,215	$7,035
Monthly Hosting Fee	None	$850

Figure 10 – Local vs. Cloud Hosted Cost Comparison

The immediate project savings is more than $40,000! You'll see in Figure 11 on page 168 that if you include the cost of maintenance and support for both options, the monthly hosting fee never catches up to that savings! In fact, the spread is widening.

The cloud provides churches with a major savings. It allows funds that would have been spent on hardware, software, and engineering to instead be invested elsewhere in ministry.

The key with cloud-hosted services is to keep your church's data private, safe, and secure. That requires some due diligence when selecting cloud-hosting vendor(s).

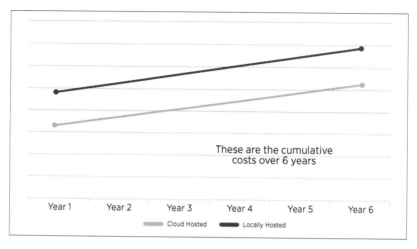

Figure 11 – Cloud vs. Locally Hosted Comparative Costs

Chapter 15 explains how to select an outsource vendor (see "Who Can You Trust?" beginning on page 121). In addition, add the following two criteria:

- *Private vs. Public Cloud.* There are two halves to the cloud: "public cloud" and "private cloud." Public cloud vendors allow anyone to create an account and begin using their services. This includes Facebook, Instagram, X (formerly Twitter), M365, and so on. Private cloud vendors require users to have been pre-approved and invited to connect to servers and services; they are not open and available to everyone.

Understanding this difference is important because churches have a lot of sensitive data. With increasing threats toward sensitive data, safeguards are essential. Placing sensitive data only in a private cloud vendor's servers is a good safeguard.

- *Datacenter Rating.* Datacenters are rated on their infrastructure redundancies: the more redundancies of power, internet trunks, and ability to manage the environment (temperature and humidity), the greater the likelihood of uptime (staying up and running). Figure 12 is a quick explanation of the ratings a church should review before selecting one.[47] Higher-numbered Tier Ratings are preferable.

Tier Rating	Redundancy	% Uptime	Max Downtime
4	at least double redundancy, a.k.a. 2N+1 fault tolerance	**99.995%**	26.3 minutes annually
3	full redundancy, a.k.a. N+1 fault tolerance	**99.982%**	1.6 hours annually
2	partial redundancy	**99.749%**	22 hours annually
1	no redundancy (only one source of power, only one internet trunk, only one way to manage HVAC)	**99.671%**	22.8 hours annually

Figure 12 – Datacenter Tier Certifications

We recommend a minimum datacenter certification of Tier 3, and prefer Tier 4. Those tiers ensure the highest levels of data availability.

---

[47] Colocation America, *Uptime Institute's Tier Standard System,* http://colocationamerica.com/data-center/tier-standards-overview.htm (last visited Feb. 6, 2024).

In addition, make certain your datacenter and/or host will keep your data in your country. John Koetsier, senior contributor for *Forbes*,[48] wrote a 2023 article detailing the 2.4 billion people restricted from internet access in just the first half of that year.[49] The list of countries included China, Russia, Brazil, Iran, India, Pakistan, Ethiopia, Mauritania, Senegal, Turkey, and more—a total of 29! Those cutoffs are usually done to reduce political unrest against a country's government with the hope of keeping pictures and stories of government actions out of the world's view.

The lesson is clear: make certain your cloud data is hosted in your country. Otherwise, your church may lose access if it resides in a country that cuts off access. If your host can't ensure this for you, find another host.

BYOD

BYOD is another popular paradigm shift. A BYOD situation is one in which a church team member uses his or her device instead of one supplied by the church. A small percentage of churches are taking advantage of its benefits, and that number is growing.

We're not sure why people would prefer to use their own devices unless they would rather work in a different operating system (Windows versus Mac or vice-versa) than the one on the device supplied by the church. However, MBS even gets requests from individuals whose devices run the same OS as that of the church-supplied device!

This trend will likely continue and grow. The cloud's ubiquitous access to data and apps makes it a viable option.

---

[48] https://www.forbes.com/sites/johnkoetsier/?sh=5c6355437d1b (last visited Feb. 6, 2024).

[49] https://www.forbes.com/sites/johnkoetsier/2023/07/19/internet-shutdowns-24-billion-people-restricted-from-internet-access-so-far-in-2023/?sh=b932c2a64cc6 (last visited Feb. 6, 2024).

BYOD makes many IT professionals uncomfortable. Historically, they have been able to keep personal devices off their networks. IT people are responsible for protecting data, so this has been the right thing to do. This policy has allowed us to protect data from malware and other threats. BYOD, though, increases threats to the data if the BYOD strategy is not implemented carefully.

However, BYOD arrangements can work for churches—and save them money. We recommend a minimum set of policies to protect the church and team members with BYOD arrangements.

Many organizations don't buy people their devices anymore. Instead, they give them a cash rebate to mitigate the upgrade costs pushed to employees.

That can save a church a lot of money and increase its team members' satisfaction! What a great way to help achieve what Clif Guy said (see page 29) about trying to have a default "yes" posture!

BYOD is a technology reality, and churches are leading the way! Some implementation methods we see churches use:

- *SYOM, Spend Your Own Money.* This is when a church gives a rebate based on a formula that considers the team member's role on the church staff and a percentage of the purchase cost with a cap.

- *CYOM, Choose Your Own Machine.* Based on team members' roles on the church staff, they get a budget that can be applied to a list of pre-approved computer choices. If the system a team member selects is beyond the established budget for that role, the department can choose to cover the additional cost out of its budget.

To make these types of arrangements work, the data needs better protection than it otherwise would need because of the increased threats that BYOD introduces. That protection should include better firewalls (don't just rely on the internet service provider's router settings) and malware protection (as well as a better disaster recovery plan (a good backup strategy)—including more depth of backup history (we recommend a minimum of one month's backups for all churches, but more when BYOD is in play).

BYOD Policies

Since BYOD requests are on the rise, it's wise for churches to research this subject and form policies before a team member asks to use his or her own device. Here is a starter list of what should be covered in a policy:

Team Member Responsibilities

- *To be productive.* Team members who request to use their personal computers and/or devices must understand that they are responsible to be productive. Thus any such BYOD request, if granted, will require that they be at least as productive as they would have been using the systems normally provided by the church. Standards of productivity are the responsibility of management, and team members who are not as productive on their own computers and/or devices will be required to use church-provided systems.

- *To be cooperative.* Personally owned computers and/or devices, if allowed to be used for work, must meet minimum standards.

Those standards will be set and modified from time to time by the IT Department or a delegated IT vendor, and may address minimum processor chipsets and operating system versions, minimum amounts of RAM and storage, and required use of specific church-provided applications, such as productivity suites (like Microsoft Office), anti-malware tools (like SentinelOne), email clients, and more. Use of substitute applications must be approved by the IT Department *and* the team member's direct supervisor.

- *To be responsible.* The team member agrees to maintain his or her personally owned computers and/or devices that have been approved for work use at a level meeting or exceeding (1) productivity levels set by management and (2) the IT Department's minimum system requirements. The team member is responsible for any costs due to failed hardware, configuration and/or software issues, and theft or breakage.

- *To protect.* The team member agrees: (1) to maintain the security of his or her personally owned computers and/or devices to protect the data and integrity of the church's systems; (2) to let his or her supervisor and the IT Department know if the device has been lost or stolen within 24 hours; and (3) to let the church install software that could delete the church's data, if the church so desires, with or without notice. The team member agrees to submit the personally owned computers and/or devices approved for work use for inspection by the IT Department when requested to confirm that the system is being properly protected against malware and other threats. The team member understands that the church may see data and files that could otherwise be considered private but agrees to hold the church harmless against any claims against loss of privacy in exchange for the church agreeing to allow the team member to use his or her personally owned computers and/or devices for work.

Church Responsibilities

- *To provide a productive environment.* The church agrees to provide a suitable work area and workstation to help the team member be productive at levels required by management. If the team member's personally owned computers and/or devices are not available due to required repairs (for which the team member is responsible), the church will provide a substitute computer or device using church-owned computers and/or devices for a reasonable period of time.

- *To be reasonably accommodating.* When a team member requests permission to use a personally owned computer or device for work, the church agrees to be reasonably accommodating if the team member can demonstrate that his or her productivity will meet or exceed the productivity standards set by the church.

- *To support.* The church is not responsible for supporting the team member's computer or device. However, the church will give help desk support at the same level as it does for church-owned computers using software provided by the church.

- *To explain exempt versus non-exempt issues.* Some team members are subject to overtime rules based on state and/or federal laws. The church is responsible to explain the team member's exempt or non-exempt status, and how that affects recordkeeping for hours worked.

Termination Procedures

If a team member is terminated by the church or resigns, the team member agrees to remove all church-owned software and data from the personally owned computer or device, or to provide it to the IT Department so the IT Department can remove the church-owned software and data.

Signed Acknowledgement

The team member and his or her supervisor will sign an agreement acknowledging the BYOD policies in place. The acknowledgement will also state that the church may modify the BYOD policy at any time and without prior notice.

Social Media

Many books and articles cover the pros and cons of social media (services like Facebook, X (formerly known as Twitter), Instagram, and many more) for churches. For this book we simply want to focus on the point that having a website alone is no longer enough if you want to attract younger generations. Young adults and young families now look for Facebook pages and YouTube channels when they're searching for a new church or researching one. It has been said that guests will visit your social media before visiting your website and before visting your building. Your social media needs to be vibrant.

Every church must adapt to this paradigm shift. Websites were important in the 1990s, 2000s, and 2010s. A social media presence is important now. That doesn't mean eliminating your church's website. Rather, it is important your church communicate via all of the "channels" your community uses: websites and social media sites.

CHAPTER 22

Artificial Intelligence (AI)

> By far, the greatest danger of Artificial Intelligence is that people conclude too early that they understand it.
>
> Eliezer Yudkowsky [50]

Artificial Intelligence (AI), like most technology tools, offers tremendous potential to help churches and ministries build The Kingdom. It can also be used in ways that would be good to avoid.

A study conducted at the University of Southern California about AI revealed surprising results in the treatment of military veterans with Post Traumatic Stress Disorder (PTSD).[51] Researchers found that veterans with PTSD were more willing to share their issues with a *virtual* human than they were with real humans![52]

---

[50] https://bernardmarr.com/28-best-quotes-about-artificial-intelligence (last visited 2/23/2024).
[51] Anna-Catherine Brigida, "A Virtual Therapist," USC Viterbi School of Engineering (Oct. 18, 2013), https://viterbi.usc.edu/news/news/2013/a-virtual-therapist.htm (last visited Feb. 6, 2024).
[52] Joseph Frankel, "PTSD Treatment: How AI Is Helping Veterans With Post-Traumatic Stress Disorder," Newsweek (Oct. 12, 2017), https://www.newsweek.com/ptsd-treatment-how-ai-could-help-veterans-post-traumatic-stress-disorder-682857 (last visited Feb. 6, 2024).

Could this technology be used in ministry? One US ministry tested AI because of the USC study and now is using it for intake bots, to generate code and database queries, and more.

Ways *Not* To Use AI in The Church

AI is a program. As such, it is written by humans and it is given sets of guidelines of what it can and cannot do. But those guidelines are set by humans.

AI does not have the indwelling of the Holy Spirit like the Bible says Christian do. Can God work through AI? Yes. Does it have the guidance of the indwelling Holy Spirit? No.

We asked ChatGPT the following question:

> In light of the Bible verses John 14:26, 1 Cor. 12:3, and 2 Cor. 5:5, give me a few paragraphs on why a pastor should or should not use AI to write a sermon.

Here's AI's five paragraph response (it's really good, and worth a full read):

> The Bible verses you've mentioned, John 14:26, 1 Corinthians 12:3, and 2 Corinthians 5:5, speak to the role of the Holy Spirit in guiding and inspiring believers. These verses emphasize the importance of spiritual discernment, the manifestation of spiritual gifts, and the assurance of God's presence through the Holy Spirit. In considering whether a pastor should use AI to write a sermon, it's essential to approach the topic with a balanced perspective, taking into account both the potential benefits and concerns.

On one hand, using AI to assist in sermon preparation can be seen as a tool that enhances a pastor's efficiency and effectiveness. AI can help analyze vast amounts of biblical and theological content, providing relevant insights and aiding in the organization of ideas. It can also assist in language refinement, ensuring clarity and coherence in conveying the intended message. This approach may free up valuable time for pastors to engage in other pastoral duties, fostering a more well-rounded ministry.

However, it's crucial for pastors to exercise caution and discernment when incorporating AI into their sermon-writing process. While AI can offer valuable assistance, it lacks the spiritual depth and inspiration that comes from a pastor's personal relationship with God and the leading of the Holy Spirit. Pastors should not rely solely on AI for the development of their sermons but rather view it as a supplementary tool. The discernment of the Holy Spirit, as mentioned in John 14:26, should still play a central role in guiding pastors as they craft their messages.

Moreover, the use of AI in sermon preparation raises ethical considerations. Sermons are not merely intellectual exercises but spiritual communications that require authenticity, sincerity, and a connection to the congregation. Pastors should be cautious not to prioritize efficiency over the personal touch and genuine spiritual guidance that comes from their own study, prayer, and relationship with God.

In conclusion, while AI can be a valuable aid in certain aspects of sermon preparation, pastors should approach its use with discernment, ensuring that their reliance on technology does not compromise the spiritual authenticity and depth of their messages. The guidance of the Holy Spirit, as emphasized in the Bible verses you mentioned, remains paramount in the life and ministry of a pastor.[53]

We asked the same question a different way that could help it justify its use in ministry settings that rely on the Holy Spirit, and AI's concluding two sentences were:

> While AI can provide valuable assistance, it cannot replace the unique insights, empathy, and spiritual intuition that a pastor brings to their role. Ultimately, the decision to use AI in sermon preparation should be guided by a pastor's commitment to maintaining the integrity of their message and the authenticity of their spiritual leadership.[54]

A Surprising Find

On Nov. 16, 2023, Nick asked AI to provide current efficiency specs on home solar panels. AI responded with, "As of my last update in January 2022...." Nick said that data was too old and requested the AI to find more current data. AI's response: "My training only includes data up to January 2022, and I don't have access to real-time information or updates beyond that date."[55]

So, before asking AI to help with any kind of data analysis, you might want to ask it how current its data is. When we asked, it said January 2022, and that it would "not be aware of events or developments that occurred after that date".[56] Keep in mind that the platform we

---

[53] https://chat.openai.com/
[54] https://chat.openai.com/
[55] https://chat.openai.com/share/a1e46957-7cd4-41f9-933b-a2f8df6b885d (Nov. 16, 2023).
[56] https://chat.openai.com/share/2a8c89b8-a4a9-4fdd-b93e-0ed1667bf7f9 (Nov. 17, 2023).

chose to use—ChatGPT—is a free AI; each AI will have different data awareness timeframes, and those that are paid versus free may as well.

Ways to Use AI in The Church

AI is a tool that, if used properly, can advance ministry. For instance, AI is great at writing first drafts of letters and providing analysis.

There is no doubt AI has been a significant technological advancement and will continue to revolutionize our lives, with some suggesting AI is to this generation what dial-up internet was to the prior generation.

But with artificial intelligence comes many questions.

- Is it safe?
- Can it be trusted?
- Will AI lead to the destruction of life on our planet?

In many ways, our sci-fi imaginations get the better of us: Is AI the Terminator come to life? Have we finally built Mr. Data from *Star Trek: The Next Generation?* When can I order my first C-3PO droid from Amazon?

Think Critically About AI

But the more constructive question for church leaders is: *How will AI affect churches and ministries?*

One thing to remember: AI has been around for a while and has been used in the form of algorithms to process data and determine outcomes. Algorithms determine our social media feeds, protect our bank accounts and personal information, and even help with traffic management.

As the algorithms get "smarter," the appearance of intelligence emerges. Add to that the ability to tackle more complex, subjective questions, such as "Which national park is the best?" and the algorithms behind AI begin to give it the appearance of discernment.

But AI can't discern. The data from which it draws its conclusions was provided by humans that may not always agree and are often filled with biases. So, while AI does its best with what it has, we've found it to be extremely flawed.

You remember the adage *garbage in equals garbage out*? It's still true. But with AI, the scale makes finding the garbage a challenge, and the subjective nature of what one human programmer views as garbage compared to another programmer's view further complicates its effectiveness.

It harkens to the early days of the internet when we emphasized that not everything you read online is true.

Now, the emphasis is on reminding people that, not only is the internet not the ultimate source of truth, but neither is social media—and neither is AI.

Embrace AI. Do Not Fear It

How does this all affect ministries?

First, there is no need to be scared. AI is not life—only God can create life. No amount of programming or algorithms can change that. AI can only mimic the creative process.

Second, you can't always trust it. Phishing scams and get-rich-quick schemes flourish because we believe what we see online. You don't know if there is another human trying to scam you or another human using AI to make the scam more complex, but you can't naively trust AI.

Third, ministries should embrace AI. (Yes, you read that correctly.)

Churches and ministries should not run and hide just because AI poses risks. Instead, they should use AI just as they use other technology: for ministry effectiveness.

Convene Conversations About AI

AI offers numerous ministry opportunities. Instead of fearing it, use it as a discipleship opportunity.

Sure, your theology will come into play when evaluating artificial intelligence, but is your church teaching about it with any theological depth?

Have you considered community events to teach the good and the bad? What about teaching basic online safety, including ways AI can be used to scam and deceive? Or why using it to cheat on one's homework or job application essays is a sin?

We assure you: we typed what you are reading here (except for the ChatGPT quote above, which we identified). But how do you know? How would a school know? Even AI tools used to detect AI-created content had to be shut down because the tools failed more than 60 percent of the time.

In many ways, AI offers an opportunity for the church to look deeper at itself, both beneath the steeple and outside the walls.

AI's Undeniable Power

Meanwhile, the power of AI is undeniable.

Its ability to generate lifelike videos is amazing. The benefits to church production in not having to record your pastor literally saying every word, but rather, setting up an AI version so you can improve efficiency is incredible. But what happens when the pastor leaves and the church retains the pastor's likeness and makes it say things the pastor would never say? Powerful technology must be applied through the lens of the Bible.

At a more individual level, what happens when artificial intelligence is used to fake the voice of one of your children calling to ask for money when your child, in reality, is safe? The video you posted on social media of your child giving a speech can also be used to get a sample of a child's voice that, in turn, could be used to scam you through emotional distress.

Churches can face similar challenges with voicemails, texts, and email messages purportedly from the pastor and asking for payments to be deposited to an account for various reasons. This again reinforces the need for churches to use strong internal controls, including in-person verifications, before authorizing any disbursements or changes in payment methods for vendors.

The world is constantly changing. We need to teach that the Bible is forever, providing a strong theological foundation so that, whatever comes next, our people are ready to handle it in a Godly manner. None of these technological developments surprised God. The Bible teaches the need to discern right from wrong, and we need to teach that same discernment when it comes to AI.

Using AI to Strengthen Ministry

Other benefits of AI for churches involve data collection and analytics. We've written about the data that churches collect and how to keep it safe, but what about using AI to better evaluate that data? Data is fine, but it's what you do with the data that really matters.

Artificial intelligence can be an ally in this effort by going through data and providing useful information to help make decisions.

AI could help close the proverbial "ministry back door" where people stop attending before leaders realize it. AI could help us better evaluate attendance patterns and changes in involvement, even comparing attendance with giving trends. This could help us understand who is at risk for leaving the church or struggling in a manner that a call or visit might prove fruitful. What was once complex and took hours can potentially be simplified and assessed in real time.

Numerous Legal Concerns

We have a long way to go to catch up with the advancements artificial intelligence has provided, and the law lags those advancements, too.

AI has quickly outdated copyright laws—and only recently has the U.S. Copyright Office begun seeking input regarding whether the creators of AI-generated content receive copyright protections.[57] Personal privacy lawsuits are just beginning, raising questions about using AI tools to generate things like prayer requests for church e-newsletters.

---

[57] U.S. Copyright Office, "Copyright Office Issues Notice of Inquiry on Copyright and Artificial Intelligence," (Aug. 30, 2023) https://www.copyright.gov/newsnet/2023/1017.html (last visited Feb. 9, 2024).

Defamation cases are being filed based often on AI-generated results created from negative content posted on websites and social media sites.

But in these cases, who's to blame? The AI? Those who programmed the AI? Or those who used the AI? The decisions and developments across all these issues will reshape how we know and understand the use of this technology even in church contexts.

Stewarding AI for Good

We are excited about AI's potential for affecting the Kingdom.

But whether with AI, social media, digital projectors, microphones, cameras, or anything else, all new technology requires responsible use.

Microphones are great, but if you don't know what you are doing, they will cause piercing feedback. Artificial Intelligence is also great, but if you don't know what you are doing, and you aren't willing to learn, the scale of the feedback could be destructive.

CHAPTER 23

Strategies for Solo Pastors

> Because God gave you your makeup and superintended every moment of your past, including all the hardship, pain, and struggles, He wants to use your words in a unique manner. No one else can speak through your vocal chords, and, equally important, no one else has your story.
>
> **Charles R. Swindoll** [58]

Many pastors serve in churches where they are the only paid staff member—or maybe one of two paid staff members. As such, they don't have the resources available to them that those who serve in larger churches possess, and their use of technology is often more of a challenge because they can't use enterprise-class strategies. In fact, their work computers are more akin to consumer-class home solutions, and they tend to be at the mercy of whoever's available to give them support when needed.

Some of that is changing, and we hope to address here some of the ways solo pastors can employ strategies to improve the reliability of their technology.

---

[58] Charles R. Swindoll, *Stones of Remembrance: Bible Study Guide* (1988).

Buying Hardware

Price is a big factor, but support is an even bigger one. Look for a vendor who can support you without forcing you to ship in your computer for repairs. Best Buy has its Geek Squad, and Apple has its Genius Bar. Look for a support service like one of these—something that will be available as conveniently as possible.

Some things to insist on when you make your purchase:

- An educational discount if you're a student or professor.

- A pro version of the operating system (you definitely do not want the "Home" or "Student" versions).

- Assistance with removing all "trialware" that comes with your computer. (The merchant often does this for a small fee, but it's worth it. Maybe you can negotiate and get it thrown in with the purchase!)

- The inclusion of extended warranty care that includes coverage for accidents involving mobile devices if, for instance, the device is a notebook or tablet.

Buying Software

Software makes it possible to get your work done. If the purchase is made by your church, you may qualify for charity pricing. If you're a student or professor, your purchase may qualify for an educational discount. Again, though, only purchase enterprise or pro versions of the software if possible. In addition, some of today's operating systems already include apps that may be adequate for word processing, spreadsheets, and so on.

Resources that Can Help

If your church is affiliated with a denomination or movement, ask if there are technology resources and solutions available to help you. Many do, and even if they come with a fee, they are worth buying.

Lastly, one final resource worth mentioning is the collection of hundreds of articles we have written over the years. Many are available on MBS' website for free! When you have an IT-related question, you may find the answer at https://www.mbsinc.com/articles/

Solo pastors are not alone! In Chapter 15 we mention four firms that focus on serving Christian churches and ministries. Contact them and ask if they have ways they can help a solo pastor.

Glossary of Terms

A

AI, Artificial Intelligence. A type of programming that allows an app to respond in context to input as though interpreting and reasoning.

Apps. The applications used on computers to accomplish tasks. Also referred to as software.

Auto Attendant. An internal telephone system's receptionist or operator that is *not* a live person, but instead is a recorded voice offering a set of menu options to the caller.

B

Backup. The practice of making copies of data as part of a disaster recovery plan.

BCP, Business Continuity Plan. A detailed plan of how an organization will continue operations during a disaster.

Bot. A usually malicious program running on the internet looking for computer and system vulnerabilities.

Botnet. A network of malware-controlled devices.

BYOD, Bring Your Own Device. Allowing organization team members to use their own computers rather than requiring them to only use those provided by the organization.

C

ChMS, Church Management Software. Database systems that help churches track congregants' contact information, contributions, attendance, and volunteer involvement. Some ChMS solutions also provide an accounting solution unique to the accounting needs of churches.

CYOM, Choose Your Own Machine. Allowing organization team members to choose their computers from a menu of options rather than requiring them to use a specific configuration.

Church IT Network. The Church IT Network, a.k.a. CITRT, is a group of church-specific IT people who gather to help each other with proven solutions and encourage each other through the challenges of managing IT in churches.

The Cloud. There are many ways to define the Cloud. The easiest definition is that the data and apps are stored and accessed online; these are often referred to as hosted data and apps.

CFO, Chief Financial Officer. The primary officer of a corporation responsible for managing the corporation's finances.

CIO, Chief Information Officer. The primary officer of a corporation responsible for managing all of the corporation's technology. This position is also often referred to as the CTO, or Chief Technology Officer.

COO, Chief Operations Officer. The primary officer of a corporation responsible for managing the corporation's day-to-day operations.

CTO, Chief Technology Officer. The primary officer of a corporation responsible for managing all of the corporation's technology. This position is also often referred to as the CIO, or Chief Information Officer.

D

Datacenter. Any place that has servers can be referred to as a datacenter. However, the term usually refers to a location that is very large and houses thousands of servers.

DDoS Attack, Distributed Denial of Service. An internet attack against computers, servers, and internet-connected devices designed to overwhelm its target with requests for attention so that its normal uses are jammed.

DRP, Disaster Recovery Plan. A detailed plan of how an organization will recover from a disaster.
DNS, Domain Name System. The worldwide system that facilitates internet connectivity by storing data about how to connect to every website, email server, and so on.

E
Encryption. A technological scrambling of data that requires a digital key to unscramble and access encrypted data. The key is provided by running an app, whether it's part of the operating system or independent of the operating system.
ESXi. VMware's hypervisor.
Ethernet. A type of cable to connect computers and other devices to the network or to the internet. Ethernet cables have modular-type plugs that "clip" in to their socket.

F
Firewall. A hardware or software solution to keep intruders out of a computer system.

G
GDPR, General Data Protection Regulation. The European Union's 2018 law requiring companies to allow someone to disappear from their databases.
GUI, Graphical User Interface. The presentation by an app of its features and data on a computer monitor or display; how a computer user interfaces with the computer app.

H
Hardware. Physical computers and other IT infrastructure devices.
Host. A server-class physical computer that has a hypervisor installed, allowing it to *host* multiple virtual (non-physical) servers.

Hosted. The location of the servers, data, and apps; hosted usually refers to an off-site location, though when located in-house it can be referred to as *locally hosted*.

HyperV. Microsoft's hypervisor.

Hypervisor. An app that allows a physical computer to be configured as a host for multiple virtual computers.

I

IGMP (Internet Group Management Protocol). A communications protocol that facilitates aggressively managed multicast data. Switches use IGMP to "listen" to the "conversation" between devices to actively manage/lighten the network load.

Infrastructure. Like roads and utilities are part of a geographic region's infrastructure, the servers, switches, and cabling that facilitate the access and sharing of data are called IT infrastructure.

Insource. Hiring employees to do tasks because the tasks are an organization's mission or core skill set.

IP, Internet Protocol. The method used to transmit data from one computer to another using those machines' IP Addresses (unique sets of numbers that identify the computer on a network or on the internet). The term IP is often coupled with another term, TCP (Transmission Control Protocol) in the format of TCP/IP.

ISO. A file that is an image of a CD or DVD that can be mounted logically just as though it was physical media. You can also burn an ISO file to optical media and run it that way.

ISP, Internet Service Provider. The company providing your internet service.

IT, Information Technology. The profession and application of computers to store, recall, transmit, and manipulate data in an organization.

J

Jumbo Frame (aka 'jumbos'). A setting allowing a packet of data to contain more than 1,500 bytes, increasing the limit as high as 9,000 bytes. Basically, this allows a *frame* of information to carry more information, referred to as the *payload*, in a single frame, improving performance of fast networks.

L

LAN, Local Area Network. A connected group of computers and devices at an organization, usually located at the organization's site.

Licensing. The terms of the legal permission that gives the purchaser of apps and hardware the right to use their purchase within certain limits and sometimes with certain restrictions.

Live Attendant. An internal telephone system's receptionist or operator that is a live person.

Local Admin. A desktop or notebook computer setting that gives the person using it the ability to make modifications, like adding an app. This capability sometimes is reserved only for the professional IT staff. Local Admin rights do not impact a user's network rights or permissions.

M

Malware. Apps that are written for the express purpose of destroying data and systems, or holding them captive for ransom.

MBS, Ministry Business Services, Inc. The IT consulting firm founded by the co-authors, Nick Nicholaou, and now owned by co-author Jonathan Smith. (*www.mbsinc.com*)

MDM, Mobile Device Management. Software designed to manage computer and device configurations and patches. See section beginning on page 163 for a more complete explanation.

MFA, Multi-Factor Authentication. The requirement of using as many as three identification verification factors when logging in to a system. This is related to 2FA (Two-Factor Authentication). See section beginning on page 151 for a more complete explanation.

Mission Critical. Something that, if it fails, will result in the failure of the organization's operations.

N

NAS, Network Attached Storage. A server-type device on a network that provides storage, acting like an external hard drive. NAS servers predate SANs, and provide file-level access to their contents.

NOS, Network Operating System. Similar to the operating system on a computer, this operating system is what makes it possible for the computer to function as a server on a network server *versus* a regular computer.

O

Outsource. Hiring non-employees to do tasks that are not an organization's mission or core skill set.

OS, Operating System. Every computer and device runs a foundational software that enables it to run apps and meet the needs of its user. Examples are Windows, Mac OSX, Android, and iOS.

P

PBX, Private Branch Exchange. Internal telephone systems typified by a central operator or receptionist.

PCI Compliance. The credit card industry's requirement for any organization processing credit card and debit card transactions.

Private Cloud. Cloud-based servers and services only accessible to those who were pre-approved to access and use them.

Productivity Software. A set of apps that usually includes word processing, spreadsheets, and other modules.

Public Cloud. Cloud-based servers and services accessible to any who would like to connect to them, such as Facebook, Dropbox, Instagram, and so on.

S

SAN, Storage Area Network. Similar to a NAS, a SAN is a server-like device that is like a large external hard drive that network users can access. It has more redundancy and safety built into it than a NAS, and is faster. SANs' contents are at the block level versus the file level, thus requiring a server or some kind of client to access the files stored on them.

Silos of Data. When team members use various databases—even spreadsheets or documents—to manage their ministries, each is a silo of data. If those silos don't have the ability to exchange and coordinate data, the church sacrifices an important synergy in accomplishing its mission.

Softphone. An app that replaces the physical telephone handset.

Software. The applications used on computers to accomplish tasks, also referred to as apps.

SSID, Service Set Identifier. The name of the WiFi signals available to connect a device to.

SYOM, Spend Your Own Money. One of the BYOD strategies in which church team members contribute to the cost of their computers or mobile devices.

T

Trialware. Software that comes on new, consumer-class computers. The software providers underwrite some of the computer cost if the manufacturer includes their solution with the computer, but the programs are just trial versions that eventually expire and must be purchased if continued use is desired. Two big problems with trialware: one, they're rarely ever worth using, and two, they don't always *cleanly* uninstall, causing future problems.

V

Virtual Host. A server-class physical computer that has a hypervisor installed, allowing it to *host* multiple virtual (non-physical) servers.

Virtualization. Computers that are not actual physical computers but are virtual configurations that act like separate computers made possible by an app referred to as a hypervisor.

VoIP, Voice Over Internet Protocol. Telephone systems that transmit calls using internet technology rather than traditional telephone technology.

VPN, Virtual Private Network. A secure method of accessing data, usually from off-site, that involves encrypting the communication signal.

Vsphere. VMware's hypervisor.

W

WAP, Wireless Access Point. The WiFi radios that broadcast WiFi signals.

X

XenServer. Citrix's hypervisor.

Miscellaneous

2FA, Two-Factor Authentication. The requirement of using as many as two identification verification factors when logging in to a system. This is related to MFA (Multi-Factor Authentication). See section beginning on page 151 for a more complete explanation.

Table of References

This is a list of people and ministries who were quoted or referenced through this book.

A
ACTS Group – *123*
Aldrin, Buzz – *41*
Apostle Paul – *97*

B
BEMA – *123*
Better Business Bureau – *123*
Bishop, Dan – *101*
Bono, Paul David Hewson – *55*
Booth, Chris – *75*
Boy Scouts – *129*
Branaugh, Matthew – *113*
Brown, David – *21*
Busby, Dan – *145*

C
Christian Camp and Conference Association – *7*
Christian Leadership Alliance – *7*
ChurchLawAndTax.com – *113*
ChurchSalary.com – *113*
Clinton, Bill – *165*
Colocation America – *169*

D
Damon, Matt – *69*
Depp, Johnny – *43*
DNS Made Easy – *128*
Drucker, Peter F. – *95, 103, 143*

E
Eisenhower, Dwight D. – *63*
Enable Resource Group – *124*

G
Gladwell, Malcolm – *8*
Guy, Clif – *29, 171*

H
Hanks, Tom – *35*
Huff, Rolla P. – *8*

J
Jethro, Moses' father-in-law – *15*
Jobs, Steve – *25, 87*

K
KnowBe4.com – *108*

L
Lawless, Chuck – *47*
LifeWay – *114*

M
Messmer, Gary – *86*
Ministry Business Services Inc. – *8*
MinistryPay – *114*

N
Norris, Chuck – *137*

P
Pearson, John – *9, 159*
Powell, Jason – *20*

R
Reagan, Ronald – *125*
Robert Half 2021 Technology Survey – *113*
Ruckus – *83, 85*
Ruth, George Herman 'Babe' – *109*

S
SonicWALL – *85, 152*
Stanley, Andy – *117*
Swindoll, Chuck – *13, 187*

T
The Church IT Network – *68, 113*
The Church Network – *68*
Twain, Mark – *81*

V
Veeam – *130, 167*

W
Wall Street Journal – *8*
Warner, Greg – *130*
Wikipedia – *16*
Willow Creek Association – *7*

About the Authors

Nick Bruce Nicholaou grew up in an agnostic home. He is the oldest of three and the only son of a couple who began their marriage very young. He chose a childhood path that destined him to an early grave, and he would have ended up there if the Lord had not gotten his attention at the age of twenty-one.

When God got a hold of his life, he gave Nick a thirst and hunger for the Scriptures and tied him to a group of believers who helped shape his desire to serve God in every way possible—how ever the Lord wanted.

After college, Nick was hired into executive service in the automobile manufacturing industry where he was tasked to help dealerships improve their customer satisfaction and success. Because he had to influence independent business owners to change some of what they did and how they did it, the Lord used that time to develop Nick's consulting skills.

Just months into his marriage, while in devotions one day, Nick sensed the Lord telling him he wanted Nick to do something different. *Probably just last night's pepperoni,* Nick thought. But then it happened again the next two days! So, Nick decided he needed to tell his wife, Grace, even though she might think him weird. She responded, "I've been getting the *same thing!*" So, they determined to discover what that meant by researching the needs of churches and how the Lord might use them to help meet those needs.

Their sense was that he wanted to take Nick's business management skills and Grace's accounting skills (she's a CPA who has focused her practice on serving Christian churches and ministries) and use them to help Christian churches and ministries across the United States. That was in 1986. They have been fulfilling that call ever since!

When Nick was in college—before the advent of the personal computer—he believed computers could be great tools to help accomplish many tasks better and more efficiently. While majoring in business administration with a management focus, he took a computer course every semester so he could gain access to the school's mainframe computer system.

Nick often says that it is amazing what God can do with someone who is willing and available to serve him, how ever He wants! And, that he, himself, is proof!

About the Authors

Jonathan Smith has been the director of technology for Faith Ministries in Lafayette, Indiana, since 2001. Jonathan oversees all technology at Faith's multiple campuses. From 2003 to 2015, Jonathan served as one of the elected deacons at Faith.

In 2021 he added to his professional responsibilities the oversight of MBS, a trusted church and ministry IT consulting and support team, by becoming its owner and president.

He is also an author and highly sought conference speaker. He wrote *#RUHooked*, a book on social media "for teens and the people who care about them" and frequently publishes articles on technology, ministry, leadership, and social media.

Before accepting the call to vocational ministry, he worked in several IT and media-related fields, including radio as an on-air personality, and as a voice-over artist for commercials and promotions for the *Star Trek* television franchise.

Jonathan attended Purdue University. He earned a bachelor's degree from the University of Phoenix in Management of Information Technology.

Jonathan enjoys traveling, eating, and watching rocket launches. Married in 1999, Jonathan and his wife Heather have two children, and resides in Lafayette, Indiana.

Made in the USA
Columbia, SC
21 June 2024